# ISO 14000
# Environmental Management

David L. Goetsch

Stanley B. Davis

Prentice
Hall

Upper Saddle River, New Jersey
Columbus, Ohio

**Library of Congress Cataloging in Publication Data**

Goetsch, David L.

ISO 14000 : environmental management / David L. Goetsch, Stanley B. Davis.

p.   cm.

Includes bibliographical references and index.

ISBN 0-13-081236-6 (case)

1. ISO 14000 Series Standards. 2. Production management—Environmental aspects.

I. Davis, Stanley    II. Title.

**Vice President and Publisher:** Dave Garza
**Editor in Chief:** Stephen Helba
**Associate Editor:** Michelle Churma
**Production Editor:** Louise N. Sette
**Design Coordinator:** Robin G. Chukes
**Cover Designer:** Mark Shumaker
**Cover art:** Mark Shumaker
**Electronic Text Management:** Marilyn Wilson Phelps, Karen L. Bretz, Melanie Ortega
**Production Manager:** Brian Fox
**Marketing Manager:** Chris Bracken

This book was set by Prentice Hall. It was printed and bound by R.R. Donnelley & Sons Company. The cover was printed by Phoenix Color Corp.

10 9 8 7 6 5 4 3 2 1
ISBN 0-13-081236-6

# Preface

## WHY WAS THIS BOOK WRITTEN AND FOR WHOM?

*ISO 14000: Environmental Management* was written in response to the need for a practical teaching resource and a how-to guide that would provide a step-by-step model for understanding the ISO 14000 standard and its supporting documents, and for implementation and registration to the standard.

It was the authors' intent in the writing of this book that it serve a dual role.

- First, it was written to be used as a primary text in courses based on ISO 14000, the international standard for environmental management systems, and as a supplemental text in courses dealing with environmental protection and environmental management.

- Second, for private and public sector organizations whose operations could have environmental impacts, the book was designed to be used as a practical hands-on manual for implementing an environmental management system conforming to ISO 14000 at an affordable cost.

Interest in ISO 14000 registration has gained momentum even faster than was the case with ISO 9000, the international quality management standard. The reasons for this are in part obvious, and in part subtle. Since the beginning of the ecological movement in the early 1960s, concern for the well-being of the planet has become a significant political force. Contamination of the atmosphere, of rivers, lakes and oceans, and of the soil was largely overlooked or ignored until recently. Given the scientific data now available to us, it is clear that the planet cannot continue to sustain its many and varied inhabitants if we continue to treat it as we have. Pollution is no longer acceptable. ISO 14000 was created by the International Organization for Standardization (ISO) with participation of 50 national delegations from around the world. Its aim is to support environmental protection and the prevention of pollution.

With the acceptance and use of ISO 9000, organizations found that by improving their processes—that is, the way they did things—not only was product or service quality improved, but also costs of production were almost always reduced. We now understand that reducing or eliminating pollution can have a similar effect. Processes that have environmental aspects can be both pollution-free and less expensive.

The initial push to reduce pollution has come from governments. That is very much in evidence in the developed nations, and is already a factor in much of the rest of the world. Just as it did with ISO 9000, a secondary emphasis is coming from large customer organizations as they flow down their environmental management requirements to their supplier tiers. Finally, the end users of products or services—customers like you and me—will eventually differentiate between producers that have good environmental track records, and those that do not.

As this unfolds, conformance to ISO 14000 is becoming a prerequisite for organizations to successfully compete in the world's markets.

## ORGANIZATION OF THIS BOOK

This book begins with a comprehensive background of the International Organization for Standardization, and of the ISO 14000 standard and why it came into being. Chapter 2 leads the reader through the component parts of ISO 14000 and their relationships with each other and with ISO 9000, and provides clarification for the sometimes confusing language of the standard. Chapter 3 establishes all of the requirements of the standard in a way that is easily understood. Chapter 4 defines and develops an understanding of the concept of an environmental management system (EMS) and its elements, structure, and supporting activities. Chapter 5 clearly defines the documentation and documentation system required, and elaborates on what must be and what need not be documented. Chapter 6 explains and clarifies the registration and audit processes, and offers recommendations for the selection of registrars. Chapter 7 expands on the concept of continual improvement as anticipated by ISO 14000. Chapter 8 provides a step-by-step process for taking the organization through preparation for registration, and finally through the registration audit—and doing it at minimum cost. Also explained is how the organization may secure many of the benefits of ISO 14000 without registration. Chapter 9 briefly discusses other related ISO standards that already exist or are anticipated. Chapter 10 offers an inclusive checklist that will assist organizations in assessing their readiness for registration or self-declaration of conformance. With the checklist, it is possible not only to determine areas that already meet the requirements of ISO 14000, but also the specific actions that must be taken for conformance or registration.

## ABOUT THE AUTHORS

David L. Goetsch is Provost of the joint campus of the University of West Florida and Okaloosa-Walton Community College in Fort Walton Beach, Florida. He also administers Florida's Center for Manufacturing Competitiveness located on the campus, and is president of the Institute for Corporate Competitiveness, a private company. Dr. Goetsch is cofounder of The Quality Institute, a partnership of the University of West Florida,

Okaloosa-Walton Community College, and the Okaloosa Economic Development Council. He currently serves on the executive board of the Institute.

Stanley B. Davis was a manufacturing executive with Harris Corporation until his retirement in 1992. He was founding managing director of The Quality Institute and is a well-known expert in the areas of implementing total quality management, statistical process control, just-in-time manufacturing, benchmarking, quality management systems, and environmental management systems. He currently serves as Professor of Quality at the Institute. He also heads his own consulting firm, Stan Davis Consulting, which is dedicated to assisting private industry and public organizations throughout North America achieve world-class performance and competitiveness.

# Contents

# CHAPTER ONE

# Background of ISO 14000 and Why It Exists

## MAJOR TOPICS

- Standards and Standardization
- International Organization for Standardization
- Origin of ISO 14000
- Objective of ISO 14000
- Scope of ISO 14000
- Applicability of ISO 14000
- Rationale for ISO 14000 Registration

## STANDARDS AND STANDARDIZATION

The subject of this book is ISO 14000, an International Standard for environmental management systems. Throughout the book we will discuss standards and standardization. Consequently, it is important that we understand the terms. Webster's defines *standard* as ". . . 3: something established by authority, custom, or general consent as a model or example . . . 4: something set up and established by authority as a rule for the measure of quantity, weight, extent, value, or quality. . . . " Further, Webster's defines *standardize* as "1: to compare with a standard 2: to bring into conformity with a standard."[1]

We use standards all the time without giving them so much as a thought; in fact, life would be difficult without them. Consider language, for instance. We are able to communicate because of standardization within a common language. Words and phrases have common (standard) meanings. Therefore, the words and phrases may be easily transferred from a speaker or writer to a listener or reader who, as a result of standardization, can understand what is being said. When there is no standard—no common meaning, as between English and Mandarin—we have great difficulty communicating our thoughts and ideas.

On another level, when you go to the grocery store to buy a package of cookies, you don't give a thought to whether a one-pound package baked in New York has the same weight as a package baked in California. Nor do you worry about whether a one-gallon milk container really holds one gallon, regardless of the dairy of origin. These things are taken for granted because we live in a society with a common set of standard weights and measures.

When you turn on a television anywhere in North America, you are confident that no matter what station you choose it will be compatible with the TV's operating parameters. This is because all broadcasters and television receiver manufacturers adhere to a set of technical standards. However, in Europe or Japan, if you tried to use the same television to receive local broadcasts, the result would be different. Why? The television standards used in Europe or Japan differ from the standards used in North America. In fact, there is no common international broadcast standard. Imagine what it would be like if every state in the United States had its own standards or, even worse, if there were none. Standards are pervasive, and if they are planned well and adopted widely, they are of great benefit to mankind.

In the business world, standardization is imperative so that a manufacturer in one country can sell its products worldwide without requiring hundreds of location–specific models. For example, a size eight dress made in Malaysia will fit just like a size eight dress made in South Carolina. Likewise, a disk manufacturer's product is compatible with any computer, regardless of brand. A system of accepted standards is essential for the widespread marketability, efficient production, and use of products—from complex mechanical devices like automobiles, and even the nuts and bolts that hold them together, to purely intellectual products such as computer programs.

Perhaps the most visible absence of a common standard involves the automobile. In much of the world cars are driven on the right side of the road, so cars are configured with driving controls and instruments on the left. In some countries, notably the United Kingdom and Japan, cars are driven on the left side of the road, so driving controls and instruments are configured on the right. This not only creates interesting situations as international travelers maneuver in traffic in rental cars, but it also makes exporting cars more expensive, since the driver layout must conform to the importing country's standard, thus requiring the manufacturer to produce both left- and right-hand drive models.

The most common kind of standard relates to some type of measurement (e.g., cycles per second, threads per inch or centimeter, dimensions, fit, shape, counts,

---

### ISO 14000 INFO

*"The fact that we have a high degree of standardization has made life simpler for us in ways so basic and so obvious that we do not even realize they exist. It has given us the free national market which we take so casually. To you as end man, the American consumer, it has given lower prices and better quality, more safety, greater availability, prompter exchange and repair service, and all the other material advantages of mass production. Is this something to be taken for granted?"[2]*
*W. Edwards Deming*

weights, and measures). Another kind of standard has to do with processes, how things are done. An example might be a quality management system conforming to the ISO 9000 standard. Another example is an environmental management system conforming to ISO 14000. These standards deal not with absolutes, but with how the quality or environmental management system is established and executed.

The subject of this book, ISO 14000, is about standardizing the approach organizations everywhere take to managing their environmental aspects and impacts. This is different from simply complying with regulatory requirements—even if the regulatory requirements were the same worldwide, which they are not.

## INTERNATIONAL ORGANIZATION FOR STANDARDIZATION

In the previous section we saw that standards are important to modern society. The more *standard* the standard is, that is, the more widely accepted and utilized, the better. Widely accepted standards lead to more efficient use of resources for producers, more equitable international competition, and lower cost to consumers. The world has too many competing standards, such as standards for electrical power generation and distribution (50 Hz versus 60 Hz), units of measure (metric versus the English system), television broadcast standards,[3] and many others. For each difference that exists between national or regional standards, some competitor is put at a disadvantage and consumers ultimately face higher costs.

The present situation is much improved from what it might have been without the worldwide movement since World War II to rationalize thousands of conflicting standards of various nations. The **International Organization for Standardization (ISO)**, based in Geneva, Switzerland, has been the standard bearer for that effort. ISO was established in 1947 to promote standards in international trade, communications, and manufacturing. ISO is a nongovernmental organization, and, contrary to widely held beliefs, is not an arm of the European Union or the United Nations. It has no power to impose its standards. ISO is comprised of representative member bodies from over 100 nations concerned with standardization. Since in most countries standardization is a function of government, nearly all of these member bodies are government organizations. The principal exception to that is the United States, whose ISO representative is the American National Standards Institute (**ANSI**), a private sector organization.

---

### ISO 14000 INFO

*ISO is not a European organization, although it is based in Geneva, Switzerland. Nor is ISO an agent of any government or federation of governments. When ISO was established in 1947, the American National Standards Institute was a founding member. ANSI is one of five permanent members of ISO's governing council and one of four permanent members of ISO's Technical Management Board. ISO is a worldwide organization.*

ISO membership falls into three categories. A **Full Member** is a national body designated by its respective country as the "most representative of standardization." Most nations are represented by full members. A nation that does not have standardization bodies may be represented by a **Correspondent Member**. A small nation with a very small economy may become a **Subscriber Member**, at a reduced membership rate. Although only full members may participate in the development of standards, ISO keeps correspondent and subscriber members informed about activities of interest. A list of ISO member bodies (full members) may be found in Appendix A.

## ORIGIN OF ISO 14000

One of the early activists for the preservation of the environment was Rachel Carson, an American marine biologist. Her landmark 1962 book, *Silent Spring*, is credited with inciting worldwide concern for ecology. During the late 1960s and early 1970s, it was becoming clear that unless the planet's ecosystems were treated more sensibly, the world would be in serious trouble. Air quality in scores of densely populated areas across the globe had already deteriorated to dangerously poor levels. Many of the earth's rivers were so polluted that they could no longer support marine life. The waters, likewise, were unsafe for human use, either for drinking or cleaning. Even the rain, normally considered to be fresh, clean, and life-giving, had in many areas turned into something that poisoned, rather than nourished, plant life, that polluted lakes, and caused acid damage to automobile finishes. Pictures transmitted from orbiting satellites clearly showed patterns of pollution across Earth. Contamination of the planet through pollution resulting from mankind's careless stewardship had become a critical issue for everyone.

| ISO 14000 INFO |
|---|

### About ANSI, the U.S. Representative to ISO

*The American National Standards Institute was founded in 1918 by five engineering societies and three government agencies. Since then it has been the administrator and coordinator of the United States voluntary standardization system. ANSI's primary goal is " . . . the enhancement of global competitiveness of U.S. business and the American quality of life by promoting and facilitating voluntary consensus standards and conformity assessment systems and promoting their integrity."[4]*

*ANSI's members include more than 1,300 United States businesses, professional societies, trade associations, and government agencies. Although the ANSI Federation includes such United States government entities as the Department of Energy, Occupational Safety and Health Administration, Department of Commerce, and the Environmental Protection Agency, these entities are conferred no special status relative to members from the private sector.*

---

**ISO 14000 INFO**

*I grew up in Augusta, Maine's capital city, during the 1940s. One of Maine's major rivers, the Kennebec, flows through the middle of town. It would have been an ideal setting for boys who liked to swim, boat, and fish. Unfortunately, none of us would even consider swimming, boating, or fishing in that river. Why? Municipal sewer systems and factories upstream had turned the river into a sludge-filled stream that looked bad and smelled worse. However, I am happy to report that today, after compliance with the pollution restrictions of the 1970s, the Kennebec is a beautiful, clear, clean river that is popular for recreational activities.*
    *Stan Davis*

---

Pollution is not new. Without a doubt, man-made environmental pollution has existed since man's first footprints appeared on the planet. However, it is probably reasonable to tie the advent of widespread damaging pollution to the Industrial Revolution. Throughout the nineteenth century and the first two-thirds of the twentieth century, factories, rapidly growing cities, electric utilities, and others—from service stations and body shops to homeowners—routinely dumped their effluents (no matter what they contained) directly into the atmosphere, into streams and rivers, or onto the soil. When there were fewer factories, when there were fewer harmful chemical compounds, and when the human population was much less, pollution was a smaller problem and was essentially ignored. In more recent years, with factories multiplying worldwide; with increased use of toxic chemical compounds in insecticides, herbicides, and fertilizers; with the capacity for individuals to create enormous amounts of pollution through their own pursuits (primarily through the use of fossil fuels); and with the growing number of sources of ecodamage, ignoring the problem is no longer an option. World population has increased from 2.5 billion in 1950 to approximately 6 billion at present. The population increase means two things: a greater propensity to pollute the environment and, at the same time, greater demands on the earth's resources and bounty. Increasing pollution and increasing natural abundance are mutually exclusive undertakings. We can make efforts to somewhat control the population although we cannot purposely reduce it. So the only variable within our control is pollution, and this we can reduce.

By the mid-1980s the voices of concern for the environment were becoming louder and more insistent. The protective layer of ozone high in the atmosphere seemed to be degrading, and at the same time the atmosphere was building up the so-called greenhouse gases that might lead to global warming. In the tropics the clearing of great swaths of the rain forest were noted, and scientists warned that the whole planet would be in peril if such irresponsible forest husbandry continued. Scientific opinions vary on the subject of ozone depletion. Some stress that continued use of chlorofluorocarbons will destroy the ozone layer. Chlorofluorocarbons, or **CFCs**, are found in common industrial solvents, air conditioning systems, and until recently in aerosol cans of hair spray, paint, and many other products. This depletion of the ozone layer, in turn, could result in widespread skin cancer.

Similarly, if we continue to burn fossil fuels (coal, petroleum products) at current or higher rates, the polar ice caps could melt and cause worldwide flooding.

Scientists have no consensus that human-induced global warming exists. One group claims the earth's temperature is cyclical over both the short and long term and that these patterns are evident today. The other group, which has captured the attention of popular media and many governments, argues that climatic change is very evident and is caused by man. Regardless of one's stand in this debate, it is certain that man has defiled the environment and the earth's ecosystems must be treated with greater respect if future generations are to have a chance for fulfilling lives.

Increasing concern for the environment did not go unnoticed by political institutions in this country and elsewhere. The U.S. National Environmental Protection Act was passed by Congress in 1969, forming the Environmental Protection Agency (EPA). The United Nations convened the Conference on the Human Environment in Stockholm in 1972. Two significant products emerged from that conference. First, the United Nations Environment Program (UNEP) was formed. UNEP was charged with the promotion of worldwide environmental responsibility and awareness. It was UNEP's job to communicate to the world that there is a problem. Second, the World Commission on Environment and Development (WCED) was established. In 1987 WCED published a report calling for industry to develop effective environmental management systems. Also in 1987, a worldwide meeting was held in Montreal to develop agreements necessary for banning production of ozone-depleting chemicals.

An outcome of the WCED report was the 1992 United Nations Conference on Environment and Development (also called Earth Summit) in Rio de Janeiro. In preparation for that conference and in recognition of the successful development of its ISO 9000 standard for quality management systems, ISO was asked to participate. During 1991 ISO with its fellow standards organization, International Electrotechnical Commission, established a Strategic Advisory Group on the Environment (SAGE) staffed by 25 volunteer countries. SAGE concluded that it would be appropriate for ISO to develop international environmental management standards and the required implementation and auditing tools. A commitment to do so was secured from ISO at the 1992 Rio de Janeiro conference.

Unfortunately, there were some early problems. To the consternation of several ISO member nations, including the United States, SAGE went beyond its mandate to rule on the need for environmental standards and actually started developing them. Serious work on environmental standards began in 1993 when ISO created Technical Committee 207 (TC 207), which it chartered to develop a uniform international environmental management system (EMS) standard and the tools necessary to implement it. Specifically excluded from TC 207's scope of work were testing methods for pollutants, setting limits on pollutants, and setting environmental performance levels. These exclusions prevented TC 207 from getting involved in work that was in the domain of regulatory bodies.

At TC 207's first meeting 22 nations were represented. Ultimately 50 national delegations participated in the development of the standard.[5] TC 207 established two subcommittees to develop the standards. Subcommittee SC 1 wrote ISO 14001 and 14004, borrowing heavily from BS 7750, but with significant input from several other nations, especially the United States. Subcommittee SC 2 wrote ISO 14010, 14011, and 14012. ISO 14000 was published in 1996.

> **ISO 14000 INFO**
>
> The *first environmental management system standard* was published in March 1992. The British Standards Institution, working with interested stakeholders from industry, government, environmental organizations, registration bodies, and consultants, developed BS 7750 along the lines of the BS 5750 and ISO 9000 quality management system standards. BS 7750 subsequently became the model for ISO 14000.
>
> Source: John Cascio, ed., The ISO 14000 Handbook (Fairfax, VA, and Milwaukee: CEEM Information Services and ASQ Press, 1996), p. 491.

## OBJECTIVE OF ISO 14000

> The overall aim of this *International Standard is to support environmental protection and prevention of pollution in balance with socioeconomic needs.*[6]

The fundamental objective of ISO 14000 is to assist organizations in preventing environmental impacts that could result from the organizations' activities, products, or services. Further, organizations adhering to ISO 14000 may be assured that their environmental performance meets and will continue to meet its legal and policy requirements. ISO 14000 attempts to do this by providing "organizations with the elements of an effective environmental management system."[7] ISO 14000 does *not* establish environmental goals or dictate absolute environmental performance requirements. Those functions are left to the organization and the regulatory agencies under which the organization operates.

## SCOPE OF ISO 14000

ISO describes the scope of ISO 14000 as follows:

> . . . specifies requirements for an environmental management system, to enable an organization to formulate a policy and objectives taking into account legislative requirements and information about significant environmental impacts. It applies to those environmental aspects which the organization can control and over which it can be expected to have an influence. It does not itself state specific environmental performance criteria.[8]

To comply with the Standard's requirements, the organization must structure its EMS so that it addresses all of the ISO 14000 clauses from Clause 4.1, General Requirements, through 4.6, Management Review. These clauses are explained in detail in Chapter 3. An organization must:

- Design an environmental management system compatible with the requirements of the standard. (4.1)

- Formulate an environmental policy. (4.2)
- List its environmental aspects (the results of activities, products, or services that can interact with the environment). (4.3.1)
- Identify all legal and other requirements applicable to its aspects. (4.3.2)
- Establish environmental objectives and targets and plan for achieving them. (4.3.3)
- Define, document, and communicate roles, responsibilities, and authorities. (4.4.1)
- Identify needs for and carry out training. (4.4.2)
- Establish and maintain procedures for internal and external communication. (4.4.3)
- Develop and make available documentation applicable to the EMS. (4.4.4)
- Provide for control and maintenance of applicable documentation. (4.4.5)
- Ensure that procedures associated with significant environmental impacts are carried out under specified conditions. (4.4.6)
- Establish and test procedures for emergency preparedness. (4.4.7)
- Monitor and measure key characteristics of operations and activities that may have significant impact on the environment. (4.5.1)
- Establish and maintain procedures to cover nonconformance and corrective and preventive action. (4.5.2)
- Establish and maintain procedures for identification, maintenance, and disposition of environmental records. (4.5.3)
- Establish a program for EMS audits to determine conformance to ISO 14000 and the EMS. (4.5.4)
- Establish an EMS management review process to ensure continuing suitability, adequacy, and effectiveness. (4.6)

## APPLICABILITY OF ISO 14000

ISO 14000 can be applied to any organization, public or private, large or small, that could have any impact on the environment and wants to:[9]

1. Implement, maintain, and improve an EMS.
2. Assure itself of conformance with its stated environmental policy.
3. Demonstrate such conformance to others.
4. Seek certification/registration of its EMS by an external organization.
5. Make a self-determination and self-declaration of conformance with this International Standard.

## RATIONALE FOR ISO 14000 REGISTRATION

No organization is forced to implement ISO 14000, at least not yet. To develop an EMS to conform with the standard will require both effort and expense. The amount of effort and expense will depend on the organization's current environmental status and disposition. Why then would an organization want to pursue ISO 14000 registration? There can be several answers to this question. Perhaps there is pressure from governmental regulatory agencies, customers, or even insurance companies. Perhaps there are concerns about liability. Perhaps profit is the motivating force. *The ISO 14000 Handbook* lists sixteen reasons for adopting the ISO 14000 EMS, including the following:[10]

- **Ease of trade**—A common International Standard, as opposed to conflicting national standards, will reduce barriers to trade.
- **Improved compliance**—The ISO 14000-certified EMS has to take into account all applicable legislative and regulatory requirements and must demonstrate the EMS's effectiveness.
- **Credibility**—If an organization is registered to ISO 14000, and periodically audited by the third-party registrar, interested parties are assured that the firm is serious about environmental concerns.
- **Reduction of liability and risk**—The firm is less apt to have environmental problems using an ISO 14000-certified EMS than a firm that is not certified.
- **Savings**—The firm will attain savings through its efforts in pollution prevention and waste reduction.
- **Favored status**—Customers favor doing business with organizations that are known to be protective of the environment.
- **Improved efficiency**—Sound, consistent environmental management methods will improve profits.
- **Pressure from stockholders**—Stockholders want to invest in companies that are environmentally proactive.
- **Pressure from environmentalists**—When environmentalists learn of a company that is not protective of the environment, they apply legal pressure to the company and its stockholders. This results in a loss of goodwill as well as litigation costs.
- **Community goodwill**—The organization's stand on environmental policy and action may be the most important factor in achieving and maintaining the community's goodwill.
- **Availability of insurance**—Insurance coverage for potential pollution incidents is more readily available, and probably less expensive, for firms that can demonstrate a system for prevention, such as that provided by ISO 14000 certification.

---

### ISO 14000 INFO

*With interest in environmental protection growing, the motivation to seek ISO 14000 registration may come down to survival. It appears that within just a few years an effective environmental management system will be the admission ticket to the game of international commerce. Without it, organizations will not be able to play.*
    *Goetsch and Davis*

Considering the state of the environment and the potential for disaster if pollution is not curbed worldwide, we believe the proper motivation for obtaining ISO 14000 registration is simply that it is the right thing to do.

## SUMMARY

1. We use standards all of the time without giving them so much as a thought. We are able to communicate because of standardization of the English language. A gallon of milk in California is the same as a gallon of milk in Florida because we have standards for weights and measures.

2. The International Organization for Standardization (ISO) based in Geneva, Switzerland, is a worldwide nongovernmental organization established in 1947 to promote common standards in international trade, communications, and manufacturing.

3. The ISO 14000 Standards had their genesis in the 1960s when worldwide concern for the environment began. In 1992 the United Nations Conference on Environment and Development (also called **Earth Summit**) took place in Rio de Janeiro. Preparation for this conference and subsequent related activities led to the development of international environmental management standards (EMS) and accompanying implementation and auditing tools being assigned to ISO. ISO created Technical Committee 207 (TC 207) to develop the standards.

4. In order to comply with ISO 14000, an organization must structure its EMS so that it addresses all of the clauses from 4.1, General Requirements, through 4.6, Management Review.

5. ISO 14000 can be applied to any organization, public or private, large or small, that might have any kind of impact on the environment.

6. Among the many reasons that justify ISO 14000 registration are the following: ease of trade, improved compliance, credibility, reduction of liability and risk, cost savings, favored status, improved efficiency, pressure from stockholders, pressure from environmentalists, community goodwill, and availability of insurance.

## KEY CONCEPTS

ANSI

Availability of insurance

CFCs

Community goodwill

Earth Summit

Ease of trade

EMS

Favored status

Improved compliance

Improved efficiency

International Organization for Standardization

Pressure from stockholders

Savings

SAGE

TC 207

WCED

## REVIEW QUESTIONS

1. Give an example of a common and widely used standard that affects your daily life.
2. Explain why standardization is important in business and industry.
3. Explain the three types of memberships available in ISO.
4. What is ANSI and what is its goal?
5. Explain *briefly* the origin of ISO 14000.
6. Describe the overall objective of ISO 14000.
7. What are the basic requirements of the following clauses from ISO 14000?
   - 4.1
   - 4.3.3
   - 4.4.2
   - 4.4.6
8. To what organizations can ISO be applied?
9. How does *credibility* promote ISO 14000 registration?
10. How does *pressure from environmentalists* promote ISO 14000 registration?

## CRITICAL-THINKING PROBLEMS

The following activities may be assigned as individual, group, or discussion activities to be completed in class or out of class.

1. Think of three different products or processes that depend on standardization, then remove the standards. What problems are going to result?
2. Most representative bodies to ISO are governmental agencies, except the U.S. representative. Does this give the United States an advantage or a disadvantage? Why?
3. Global warming, ozone depletion, overpopulation, air and water pollution—some people believe that these are critical global problems; others believe they are problems invented by politically minded groups for their own interests. What do you think and why?
4. ISO 14000 is voluntary, at least for now. Some people believe that it will have little or no impact unless it becomes mandatory; others believe that more government mandates are the last thing we need. What do you think and why?
5. Defend or refute the following statement and explain your reasons: "Every organization with fifty or more employees should pursue ISO 14000 registration immediately."
6. Explain why Apple Computer has been relegated to a niche market.
7. List five products that you think would benefit from standardization.

===== DISCUSSION CASE =====

## Environmentally Sound Hotel

"The Rittenhouse Regency Hotel in Philadelphia is being renovated into the first environmentally smart hotel in the continental United States. The hotel will provide fresh filtered air 24 hours a day to each hotel room independent of the heating and cooling systems in order to maintain energy efficiency without negatively affecting indoor air quality. State-of-the-art environmental and sustainable building materials are being used, including paint, furniture, and carpeting with low or zero toxic emissions. The hotel's staff will use nontoxic, environmentally safe cleaning and laundry products, and receive training in environmental maintenance techniques."[11]

### Discussion Questions

Discuss the following questions in class or out of class with your fellow students:

1. Is this effort of the Rittenhouse good business or just a gimmick? Why?
2. Do you think other hotels will follow the example of the Rittenhouse? Why or why not?

===== ENDNOTES =====

1. *Merriam Webster's Collegiate Dictionary,* 10th ed. (Springfield, MA: Merriam-Webster, Incorporated, 1993), pp. 1145-1146.
2. Dr. W. Edwards Deming, *Quality, Productivity, and Competitive Position* (Cambridge, MA: Massachusetts Institute of Technology, Center for Advanced Engineering Study, 1982), p. 345.
3. Standards related to electrical and electronic engineering are handled by another organization, also based in Geneva, the International Electrotechnical Commission (IEC).
4. American National Standards Institute.
5. Tom Tibor, *ISO 14000: A Guide to the New Environmental Management Standards* (Chicago: Irwin Professional Publishers, 1996), p. 17.
6. American National Standards Institute, *Environmental Management Systems— Specification with Guidance for Use* (Milwaukee: ANSI/ISO 14001-1996), p. vi.
7. Ibid.
8. Ibid., Clause 1, p. 1.
9. Ibid.
10. John Cascio, ed., *The ISO 14000 Handbook* (Fairfax, VA, and Milwaukee: CEEM Information Services and ASQ Press, 1996), pp. 10-11.
11. Kristie Guillotte, "Hotel Provides Environmentally Sound Stay," *Environmental Protection*, vol. 9, no. 5, May 1998, p. 14.

# CHAPTER TWO

# Decoding ISO 14000

- Component Parts of ISO 14000 and Their Relationships
- Language of ISO 14000
- Legal Considerations and Requirements

## COMPONENT PARTS OF ISO 14000 AND THEIR RELATIONSHIPS

Technical Committee 207 was formed by ISO to develop ISO 14000. The new environmental standard, like the ISO 9000 Quality Standard, was aimed at management systems rather than absolute specifications of performance. Therefore, ISO concluded that ISO 14000 could be structured similar to ISO 9000. TC 207 and TC 176, the technical committee responsible for ISO 9000, were directed to work together to make use of the lessons learned during the development and implementation of ISO 9000 and to build on its foundation. Feedback from the ISO 9000 user community also benefited ISO 14000 and made it easier to decode.

ISO wanted the new standard to be familiar in structure and philosophy so that users of ISO 9000 could build on their quality management systems to encompass both ISO 9000 and 14000. This may turn out to be visionary because ultimately these two standards, with the addition of a yet-to-be-developed standard on occupational health and safety, could logically be merged. In fact, designing them in a way that would facilitate merging is logical. However, much work remains before harmonization of the two standards is a reality. ISO is promoting harmonization for future revisions, to the point of insisting that the revision schedules of the two standards be synchronized. The Technical Management Board of ISO considers revision synchronization to be "imperative in order to ensure as much compatibility as possible and to facilitate joint auditing."[1]

The ISO 9000 series of standards was published as fourteen documents in its 1994 revision. The actual standards, ISO 9001, 9002, and 9003, define the requirements of the standard. The ISO 9000-1, 9000-2, 9000-3, and 9000-4 documents are guidelines for expanding on and clarifying the actual standards. In addition, four more guideline documents provide further clarification. The auditing side of ISO 9000 uses three more documents, ISO 10011-1, 10011-2, and 10011-3. In contrast, ISO 14000 currently has just five corresponding documents:

- ISO 14001 Environmental Management Systems—Specification with Guidance for Use is the equivalent of ISO 9000 standards (ISO 9001, 9002, and 9003) and seven guidelines documents. Unlike ISO 9000, ISO 14001 includes its own clarification in the form of an annex within the single specification document.
- ISO 14004 Environmental Management Systems—General Guidelines on Principles, Systems and Supporting Techniques is the only stand-alone guidelines document related to ISO 14001. It is the equivalent of all eight ISO 9000 guidelines documents.
- ISO 14010 Guidelines for Environmental Auditing—General Principles
- ISO 14011 Guidelines for Environmental Auditing—Audit Procedures—Auditing of Environmental Management Systems
- ISO 14012 Guidelines for Environmental Auditing—Qualification Criteria for Environmental Auditors

The relationship of the ISO 9000 and ISO 14000 documents is shown in Figure 2-1. Since quality management systems and environmental management systems are not

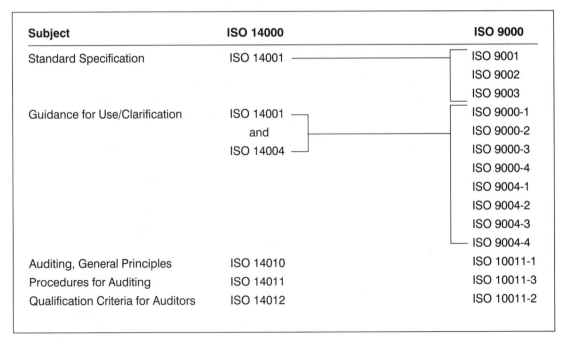

| Subject | ISO 14000 | ISO 9000 |
|---|---|---|
| Standard Specification | ISO 14001 | ISO 9001 |
| | | ISO 9002 |
| | | ISO 9003 |
| Guidance for Use/Clarification | ISO 14001 and ISO 14004 | ISO 9000-1 |
| | | ISO 9000-2 |
| | | ISO 9000-3 |
| | | ISO 9000-4 |
| | | ISO 9004-1 |
| | | ISO 9004-2 |
| | | ISO 9004-3 |
| | | ISO 9004-4 |
| Auditing, General Principles | ISO 14010 | ISO 10011-1 |
| Procedures for Auditing | ISO 14011 | ISO 10011-3 |
| Qualification Criteria for Auditors | ISO 14012 | ISO 10011-2 |

**Figure 2-1**
Relationship Between ISO 14000 and ISO 9000 Documents

exactly the same, there is no one-to-one correspondence for the standards between the documents. Figure 2-1 relates the most similar documents.

Figure 2-2 illustrates the interrelationships of the five ISO 14000 documents with which we are concerned.

## ISO 14001 Environmental Management Systems—Specifications with Guidance for Use

This specification for an environmental management system (EMS) specifies the EMS elements that must be satisfactorily addressed and implemented for registration. All the requirements for ISO 14000 registration are in this document. They are explained in chapter 3.

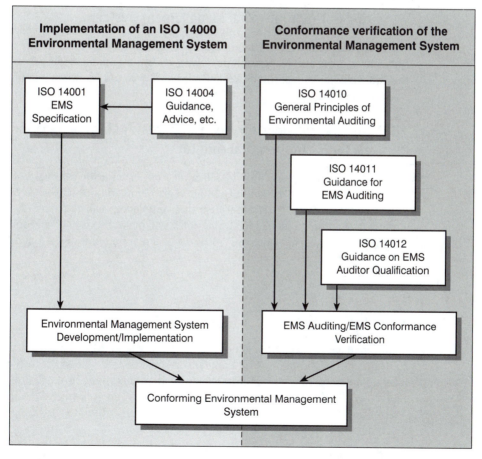

**Figure 2-2**
ISO 14000 Document Interrelationship

## ISO 14004 Environmental Management Systems—General Guidelines on Principles, Systems and Supporting Techniques

This informational document offers advice, examples, and options for developing and implementing an EMS and for integrating it into the existing management system. ISO 14004 is explained in chapter 3.

## ISO 14010 Guidelines for Environmental Auditing—General Principles

As the name implies, this document is aimed at "environmental auditing," not just the auditing of an EMS. One could question why the developing subcommittee, SC 2, did not use the more focused definition of an EMS audit given in ISO 14001; however, the principles listed are appropriate. ISO 14010 is explained in chapter 6.

## ISO 14011 Guidelines for Environmental Auditing—Audit Procedures— Auditing of Environmental Management Systems

This document provides guidance on the elements of an EMS audit, including audit objectives, roles and responsibilities of audit team members, audit preparation, conduct of the audit, audit report, and related topics. The various components of ISO 14011 are explained in chapter 6.

## ISO 14012 Guidelines for Environmental Auditing—Qualification Criteria for Environmental Auditors

This guidance document provides qualification criteria for external and internal EMS auditors. Many firms seeking ISO 14001 registration will not be able to comply with the auditor qualification criteria because their employees will not have the experience or training. For that reason, ISO 14001 does not make ISO 14012 a requirement. Registrars, however, will consider ISO 14012 to be a requirement for their auditors (external). This subject is explained in chapter 6.

# LANGUAGE OF ISO 14000

The language of ISO 14000 is not as obscure as that of ISO 9000. Even so, meanings of selected terms and phrases that apply to it must be understood. Each of the ISO 14000 documents contains definitions related to that document in its Clause 3, except ISO 14010 where definitions are in Clause 2. For clarity, the following definitions of selected terms appear in alphabetical order, not in ISO document sequence. Study the definitions in order to understand the language of ISO 14000.

*Audit Conclusion (ISO 14010, Clause 2.1)* **Audit conclusion** is the "professional judgment or opinion expressed by an auditor about the subject matter of the audit, based on and limited to reasoning the auditor has applied to audit findings."

*Audit Evidence (ISO 14010, Clause 2.3)* **Audit evidence** is verifiable information, records, or statements of fact used by the auditor to determine if audit criteria are met. It may be qualitative or quantitative and is typically based on interviews, documents, observation of activities and conditions, and existing measurement or test results.

*Audit Findings (ISO 14010, Clause 2.4)* **Audit findings** are the "results of the evaluation of the collected audit evidence compared with the agreed audit criteria." They form the basis for the audit report.

*Client (ISO 14010, Clause 2.8)* **Client** is the "organization commissioning the audit." One would normally expect the commissioning organization to be the firm under audit, which is most often the case, but not always. The client may be a second party, such as a prime contractor customer, for example. Also, since environmental audits pose some legal risk—for example, finding that a regulatory requirement is not being met—the client may be a law firm retained by the company being audited. Some organizations believe that the resulting attorney-client privilege may offer them privacy and protection.

*Conformance versus Compliance* It is important to understand and distinguish between the ISO 14000 use of the following terms, although they are not specifically defined in the standard. This understanding is especially true if the reader is familiar with ISO 9000. The words *conform, comply, conformance,* and *compliance* are similar. Webster's defines them for our context as follows:

> *Comply:* To act in accordance with a request, demand, order, rule, etc."2
> *Conform:* To bring into harmony or agreement."3

These terms are used differently between ISO 9000 and ISO 14000. ISO 9000 generally uses the term *conform* in association with products or materials. *Comply* is used in connection with meeting, or failing to meet, the requirements of the quality management system or the standard, as in requiring internal audits to ensure that " . . . quality activities and related results *comply* with planned arrangements. . . . "4 ISO 14000, however, reserves *comply* and **compliance** for governmental regulatory requirements. (ISO 9000, being a Quality Standard, faced no such mandatory requirements from regulatory agencies.) *Conform* and **conformance** are used in connection with requirements of ISO 14000 and the organization's EMS. A firm might *comply* with ISO 9000 but will *conform* to ISO 14000—but only if it also *complies* with applicable regulatory requirements. The definitions given earlier may be restated for ISO 14000 as follows:

> *Comply:* To act in accordance with all applicable regulatory requirements.
> *Conform:* To bring the environmental management system into harmony or agreement with the requirements of ISO 14001.

*Continual Improvement (ISO 14001, Clause 3.1)* **Continual improvement** is defined as a "process of enhancing the environmental management system to achieve improvements in overall environmental performance in line with the organization's environmental policy." In other words, satisfying ISO 14001 and applicable regulatory requirements may be sufficient for initial registration, but to maintain registration the organization will have to demonstrate that it is continually improving environmental performance.

*Environment (ISO 14001, Clause 3.2)*   Environment is "Surroundings in which an organization operates, including air, water, land, natural resources, flora, fauna, humans, and their interrelation. Surroundings . . . extend from within an organization to the global system."

*Environmental Aspect (ISO 14001, Clause 3.3)*   An environmental aspect is any "element of an organization's activities, products or services that can interact with the environment." In simple terms an environmental aspect is anything resulting from the organization's activities, products, or services that has the *potential* to cause an environmental impact, even if it is presently controlled to prevent such impact. The fact that the potential exists (if something goes wrong, for instance) makes it an environmental aspect. Possible environmental aspects include but are not limited to the following:

- Emissions to the atmosphere
- Discharges to water or soil
- Generation of waste
- Use of natural resources
- Community impact
- Generation of noise, dust, odors, etc.

Environmental aspects also can be positive, for example:

- Decontaminating soil
- Removing pollutants from air or water
- Recycling used materials
- Restoring land, flora, or fauna

For the purposes of ISO 14000, an environmental aspect is anything resulting from the organization's activities, products, or services that might result in an environmental impact, good or bad.

*Environmental Impact (ISO 14001, Clause 3.4)*   Environmental impact is "any change to the environment, whether adverse or beneficial, wholly or partially resulting from an organization's activities, products or services." If an environmental aspect of an organization is "emissions to the atmosphere," a possible *environmental impact* is pollution of the air. Thus, an environmental impact is the result of an environmental aspect.

*Environmental Management System (ISO 14001, Clause 3.5)*   Environmental management system (EMS) is "the part of the overall management system that includes organizational structure, planning activities, responsibilities, practices, procedures, processes and resources for developing, implementing, achieving, reviewing and maintaining the environmental policy." The EMS must include an organizational chart and be staffed accordingly; it requires planning on how the organization will deal with environmental issues; it defines who is responsible for all environmentally related activities including processes and aspects; it requires the development and deployment of environmental practices, procedures, and processes; and it must have the necessary capital and human resources available. This is the same concept as the ISO 9000 quality management system, except it is oriented to environmental concerns.

*Environmental Management System Audit (ISO 14001, Clause 3.6)*   **Environmental management system audit** is defined as "a systematic and documented verification process of objectively obtaining and evaluating evidence to determine whether an organization's environmental management system conforms to the environmental management system audit criteria set by the organization, and for communication of the results to management." ISO 14010, Clause 2.9, gives a definition that is somewhat broader, aiming at "environmental audit" rather than "environmental management system audit." The ISO 14010 version pertains to external audits only and specifies that "the results of this process" be communicated to the client, not to management. The definition of ISO 14001, Clause 3.6, is the one to which an ISO 14000-registered organization will be held.

It is important to understand that the organization develops its own EMS to meet the requirements of ISO 14001. It also sets the audit criteria, again to satisfy the standard. Then the audit determines whether the EMS conforms to the criteria, and regardless of whether it is an internal or external audit, the results are forwarded to the organization's management. (If an external audit is requested by a client, not the organization itself, the results are forwarded to the client.)

*Environmental Management System Audit Criteria (ISO 14011, Clause 3.3)*   **Environmental management system audit criteria** are defined as "policies, practices, procedures or requirements, such as those covered by ISO 14001 and, if applicable, any additional EMS requirements against which the auditor compares collected evidence about the organization's environmental management system." The organization defines the audit criteria through documented policies, practices, procedures, and requirements. The organization also identifies any applicable regulatory requirements, and these become part of the audit criteria. This definition also applies to the generic term *audit criteria*.

*Environmental Objective (ISO 14001, Clause 3.7)*   **Environmental objective** is an "overall environmental goal, arising from the environmental policy, that an organization sets for itself to achieve, and which is quantified where practicable." The organization develops its own environmental policy, and from that it establishes broad environmental objectives toward which it strives. These objectives are strategic, in that they define "how the environmental policy will be achieved." (See the definition of Environmental Target.)

*Environmental Performance (ISO 14001, Clause 3.8)*   **Environmental performance** is the "measurable results of the environmental management system, related to an organization's control of its environmental aspects, based on its environmental policy, objectives and targets." In other words, it is how well the organization is controlling its environmental aspects relative to what it had planned.

*Environmental Policy (ISO 14001, Clause 3.9)*   **Environmental policy** is the "statement by the organization of its intentions and principles in relation to its overall environmental performance which provides a framework for action and for the setting of its environmental objectives and targets." This policy will be the highest level document in the organization's hierarchy of documents that explains what it intends to do through its EMS. (See the definitions of Environmental Objective and Environmental Targets.)

*Environmental Targets (ISO 14001, Clause 3.10)*   **Environmental targets** is a "detailed performance requirement, quantified where practicable, applicable to the organization or parts thereof, that arises from the environmental objectives and that needs to be set and met in order to achieve those objectives." These are the tactical objectives that define specifically what must be accomplished to satisfy the environmental objectives (see the definition of Environmental Objectives) and, thereby, the environmental policy. These targets may require participation of the entire organization or selected parts of it.

*Interested Party (ISO 14001, Clause 3.11)*   **Interested party** is an "individual or group concerned with or affected by the environmental performance of an organization." It can include any stakeholder of the organization or the community in which it operates, a governmental agency, a user of the product or service, a customer, or environmental organizations such as Sierra Club or Greenpeace.

*Organization (ISO 14001, Clause 3.12)*   **Organization** is a "company, corporation, firm, enterprise, authority or institution, a part or combination thereof, whether incorporated or not, public or private, that has its own functions and administration. For organizations with more than one operating unit, a single operating unit may be defined as an organization." *Organization* applies equally to public and private entities.

*Prevention of Pollution (ISO 14001, Clause 3.13)*   **Prevention of pollution** is the "use of processes, practices, materials or products that avoid, reduce or control pollution, which may include recycling, treatment, process changes, control mechanisms, efficient use of resources and material substitution." The definition includes practices that declare the intent of pollution prevention, processes that may have to be altered for prevention, use of different materials, and other related practices.

## LEGAL CONSIDERATIONS AND REQUIREMENTS

ISO 14000 does not prescribe environmental performance, nor does it dictate goals for pollution prevention. Additionally, no organization is forced to register to ISO 14000. These statements seem to imply that legal implications for organizations and individuals who become involved with ISO 14000 should be few or none, just as there are few such implications with ISO 9000. However, one significant difference with ISO 14000 changes the situation: Regulatory bodies in many parts of the world have developed environmental regulations according to which organizations within their jurisdiction must operate.

In the United States, the Environmental Protection Agency (EPA) was established by Congress to protect human health and the environment. Congress granted the EPA full authority to establish and enforce environmental regulations. Congress itself has enacted laws, such as the Clean Water Act and the Clean Air Act, restricting the venting or discharging of polluting effluents. In addition, the states have created and empowered their own versions of the EPA. State regulatory bodies are free to set their own pollution limits so long as they are "at least as, or more stringent than" national regulations and statutes.

All private and public organizations in the United States must operate within the pollution parameters set by statutes, the EPA, and the various state regulatory agencies. Consequently, even without ISO 14000, organizations which in the course of their operations have the potential to affect the environment are legally obligated to comply with all applicable legislation and regulations. Penalties for failing to do so can be severe.

Regulations in the United States and most other developed countries cover chemicals, hazardous substances, gaseous emissions, wastewater effluents, and solid/hazardous waste, even noise, vibration, and odor. Coverage is not limited to what we would normally think of as dangerous compounds and poisonous emissions; everyday items such as paint and petroleum products are also subject to strict controls. Paint and body shops and neighborhood service stations must comply with environmental regulations to the same extent as Exxon, Mobil Corporation, DuPont, or any other larger company.

## Legal Considerations

ISO 14000 is designed to help organizations manage their responses to all applicable regulatory concerns through an effective environmental management system. In order to conform with ISO 14000, an organization must meet the standard's requirements for an EMS and comply with all applicable government regulations. This means that when an organization is audited by the ISO 14000 registrar, regulatory compliance is part of the audit.

It is possible that auditors could find themselves facing a dilemma with legal dimensions. If the auditor (or registrar) finds that the organization meets regulatory requirements, but a regulatory agency finds the contrary, is the auditor liable? Suppose the auditor finds that the organization does not comply with a given environmental regulation. Is the auditor obliged to disclose this fact to regulatory agencies? This is one of the reasons that law firms are sometimes the official client for the ISO 14000 registrar. Some people consider that a law firm representing the organization as an intermediary client for audits is a way to mitigate legal exposure for the auditor, registrar, and the organization. Attorney-client privilege in such cases is used to maintain the confidentiality of the auditor's findings.

## Legal Requirements

ISO 14001, Clause 4.3.2, requires the organization to have and use a procedure for identifying all applicable statutory and regulatory requirements. It further stipulates that these legal and other requirements be accessible. ISO 14004, Clause 4.2.3, states that the

---

**ISO 14000 INFO**

*The intent of ISO 14000 is not only that participating organizations meet the requirements of the legal and regulatory agencies. It also means that the organizations, through their environmental management systems and adherence to requirements of the Standard, perform to a higher and continuously improving level of environmental protection.*

legal requirements should be communicated to and understood by employees involved in the affected environmental aspects.

The ISO 14001 clause is the one that requires the organization to define and document legal regulations, permits, licenses, and codes under which it operates. This recording becomes part of the EMS documentation and, as such, is audited. It is not the auditor's or registrar's responsibility to determine which regulations apply. The intent of Clause 4.3.2 is that the organization have a way of determining which codes and regulations apply and ensuring that affected employees are informed and understand the regulatory requirements. It is further intended that the regulations be readily available to any employee who might require them. This subject is developed more fully in chapter 3.

## SUMMARY

1. ISO 14000 currently consists of five documents: ISO 14001 Environmental Management Systems—Specification with Guidance for Use; ISO 14004 Environmental Management Systems—General Guidelines on Principles, Systems and Supporting Techniques; ISO 14010 Guidelines for Environmental Auditing—General Principles; ISO 14011 Guidelines for Environmental Auditing—Audit Procedures—Auditing of Environmental Management Systems; and ISO 14012 Guidelines for Environmental Auditing—Qualification Criteria for Environmental Auditors.

2. Each of the documents that collectively comprise ISO 14000 has a section for definitions related to that document. Especially important terms from these various documents include: audit conclusion, audit evidence, audit findings, client, conformance versus compliance, continual improvement, environment, environmental aspect, environmental impact, environmental management system, environmental management system audit, environmental management system audit criteria, environmental objective, environmental performance, environmental policy, environmental targets, interested party, organization, and prevention of pollution.

3. ISO 14000 does not prescribe environmental performance, and no organization is forced to register to ISO 14000. Environmental performance is the domain of state and federal regulatory agencies. However, some concern exists in business and industry that ISO 14000 audits might reveal regulatory noncompliance. For this reason some organizations ask an attorney to serve as an intermediary client, in the hope that the attorney-client privilege will protect the confidentiality of audit findings.

4. ISO 14001 requires organizations to have and use a procedure for identifying all applicable statutory and regulatory requirements. It also stipulates that legal and regulatory requirements be communicated to and understood by employees involved with the environmental aspects.

## KEY CONCEPTS

Audit conclusion                    Audit findings
Audit evidence                      Client

| | |
|---|---|
| Compliance | Environmental objective |
| Conformance | Environmental performance |
| Continual improvement | improvement |
| EMS audit | Environmental policy |
| EMS audit criteria | Environmental targets |
| Environmental impact | Interested party |
| Environmental improvement | Organization |
| Environmental management system (EMS) | Prevention of pollution |

## REVIEW QUESTIONS

1. List and briefly explain each of the five documents that comprise ISO 14000.
2. Define the following terms:
   - Audit conclusion
   - Client
   - Environment
   - Environmental management system
   - Organization
3. Distinguish between *conformance* and *compliance* as they relate to ISO 14000.
4. Distinguish between *environmental aspect* and *environmental impact*.
5. Explain how some organizations that seek ISO 14000 registration attempt to limit their legal and regulatory exposure.

## CRITICAL-THINKING PROBLEMS

The following activities may be assigned as individual, group, or discussion activities to be completed in class or out of class.

1. Your boss, who is familiar with ISO 9000, has asked you to prepare one or two paragraphs that summarize the similarities of ISO 9000 and ISO 14000. Write the paragraph(s).
2. Identify and describe three different records (or other types of information) that might be used as audit evidence.
3. Select a company or organization with which you are familiar or to which you have access. Identify five environmental aspects in that company or organization.
4. Assume that you are a manager in Company XYZ. You have decided to propose to your colleagues that the company adopt a set of environmental objectives. Draft three such objectives.
5. Manager A says, "I am recommending that we stay away from ISO 14000. All it will do is cause us legal headaches." Manager B responds, "Maybe. But the legal prob-

lems you are worried about are nothing compared to the problems we will have if we don't get focused on solving our environmental problems." Who is right in this debate and why? What are your recommendations in this case?

## DISCUSSION CASE

"As you know, we have had our problems with the Environmental Protection Agency and the community about our painting processes," said George Hemphill, CEO at Metal Containers, Inc. (MCI). "I have encouraged the government for years to let us use a less toxic or nontoxic paint. Unfortunately, once a material—in our case, this paint—becomes part of an official government specification, changing it is almost impossible. Our only option, as I see it, is to find better ways to manage our painting processes."

"George, are you proposing that we pursue ISO 14000 registration?" asked Morris Markum, Chairman of the Board.

"I'm not ready to recommend it, but I would like us to discuss it," said Hemphill. "I don't think we can put off this discussion any longer."

As a class, play the role of MCI's board of directors, and hold the discussion they have been postponing. Debate all of the pros and cons.

## ENDNOTES

1. ISO Press Release, 9/16/98.

2. *Webster's New World Dictionary of the American Language* (New York: The World Publishing Company, 1962), p. 300.

3. Ibid., p. 308.

4. ANSI/ASQC Q9001-1994, Clause 4.17.

# Requirements of ISO 14000

- Relationship of ISO 14000 and Regulatory Requirements
- Specific ISO 14000 Requirements

## RELATIONSHIP OF ISO 14000 AND REGULATORY REQUIREMENTS

National legislative bodies and regulatory agencies of the world—not ISO—develop environmental laws and regulations. ISO 14000 requires only that registered organizations commit to compliance with those laws and regulations and that they manage their compliance through a structured EMS that conforms to the ISO 14001 standard. Governments and their agencies are the only entities that decide, for example, how much nitrous oxide or sulfur dioxide may be emitted in their jurisdictions. ISO 14000 firms must commit to meet those requirements and employ an EMS which satisfies ISO 14000. ISO 14000 firms in different localities are, in fact, required to meet widely different environmental requirements. Nonetheless, they all must approach complying using the same ISO 14000 EMS that is periodically audited by third-party registrars.

Since ISO 14000 requires the use of **continual improvement** techniques, registered firms ultimately will outperform those that seek merely to satisfy legal and regulatory requirements without ISO affiliation. Both the registered and nonregistered firms must

| ISO 14000 INFO |
| --- |
| " . . . this International Standard does not establish absolute requirements for environmental performance *beyond **commitment**, in the policy, to compliance with applicable legislation and regulations . . .* "[1] |

start at the same point (i.e., meeting the minimal legal and regulatory requirements). However, while the nonregistered firm may be satisfied with maintaining that performance level, the ISO 14000 firm *must* seek to continually improve its performance. Continually improving performance can result in lower pollution emissions, lower costs, lower utilization of resources, and other benefits. In addition to improving the environment, this will lead to the ISO 14000 firm achieving a competitive edge over the nonregistered firm.

## SPECIFIC ISO 14000 REQUIREMENTS

ISO 14001, Clause 4, contains the ISO 14000 requirements for an EMS. The requirements are divided into these categories:

| | |
|---|---|
| 4.1 General requirements | 4.4 Implementation and operation |
| 4.2 Environmental policy | 4.5 Checking and corrective action |
| 4.3 Planning | 4.6 Management review |

The first category, 4.1, requires the organization to have an EMS. The other five categories correspond, respectively, to ISO 14000 principles and elements that define an EMS: commitment and policy, **planning**, implementation, measurement and evaluation, and review and improvement. Each category is explained in this chapter, beginning with an overview, then the ISO 14001 clause, followed by requirements of the clause, and, finally, information on applying the requirements to the organization.

### Requirements Section 1: General—Environmental Management System

---

### ISO 14001, Clause 4.1: General Requirements

The organization shall establish and maintain an environmental management system, the requirements of which are described in the whole of clause 4.

---

Requirement—Clause 4.1

This is the first of many instances of the use of *shall* in ISO 14001. Wherever the word *shall* appears, it signifies a requirement. In this case the clause requires a registered organization, or one seeking registration, to develop for itself and continually use an EMS following the model outlined in ISO 14000. The remaining clauses under Section 4 detail the specific requirements for the EMS.

Application Information—Clause 4.1

Clause 4.1 is a broad, overall requirement for an EMS. It may be wise to keep the EMS as simple as possible while conforming to the standard's requirements. ISO permits a good

---

**ISO 14000 INFO**

*"The successful implementation of an environmental management system calls for the commitment of all employees of the organization. . . . This commitment should begin at the highest levels of management."*
   *ISO 14001, A.4.1*

---

deal of latitude. While the organization's EMS must address all ISO 14001 requirements, the EMS may be as simple or as complex as an organization wants to make it. No bonus points are awarded for going beyond what is required by the clauses. Keep in mind that this is a system for managing the organization's environmental functions, not the amount or severity of environmental impact. The best approach is to:

■ Maximize the use of what the organization already has. For example, if the organization is already registered to ISO 9000, it can use much of the existing quality management system directly, and more can be converted to the EMS. If you have effective procedures in place, use them. You may have to reformat them, but this will be easier than starting from scratch. Whatever you have in place that is applicable and effective should be used; it is not necessary to start over.

■ Keep the EMS as simple as possible. The system will work better when it is easy to understand and easy to follow. Simple procedures are also easier to assess, and they curtail misinterpretation by employees or auditors. If the organization later determines that a procedure needs to be expanded, it can do so, making certain to inform the registrar of the change.

## Requirements Section 2: Commitment and Policy

This clause with its subclauses reflects the first of five EMS principles listed in ISO 14004. ISO 14004, Clause 4. Principle 1, Commitment and Policy, states, "An organization should define its policy and ensure commitment to its EMS." ISO rightly believes that without the commitment of top management, the organization cannot and will not meet the ISO 14001 requirements. ISO stresses the need for commitment by top management because implementing ISO 14000 will require the participation and support of every element of the organization. Simply delegating its implementation to a group whose primary responsibility may be environmentally oriented is not sufficient.

Engineering personnel may be asked to develop new processes or products that have less propensity to pollute. Manufacturing personnel will be expected to follow approved procedures exactly. Employees in these departments must be trained in environmental awareness and proper handling and use of hazardous materials. Human resources personnel may be required to hire individuals with expertise in environmental protection. Money will be required for training, development of processes and procedures, and for environmental technology. Additionally, internal environmental auditors will be needed to audit the EMS performance of all affected departments. The only way that all these things can happen is for top management to be directly, personally, and actively involved in the EMS. If this level of commitment is missing, it will not be possible to gain and sustain the needed interdepartmental cooperation. Years of experience worldwide with Total Quality Management and ISO 9000 (both place demands similar to ISO 14000 on an organization) have proven that if top management is not totally committed to the adoption of a standard, such as ISO 14000, it will not happen.

---

## ISO 14001, Clause 4.2: Environmental Policy

Top management shall define the organization's environmental policy and ensure that it.... [continued in 4.2 a)]

---

### Requirement—Clause 4.2

As a part of the organization's EMS, an **environmental policy** *shall* (must) *be defined* by the organization's *top management*. ISO 14001, Clause 3.9, explains that the environmental policy is a statement of the organization's intentions and principles related to its environmental performance. Clause 4.2 further requires top management to ensure that the requirements of Subclauses a) through f) are satisfied.

### Application Information—Clause 4.2

Clause 4.2 in its statement "Top management shall define...." does not mean the top managers must actually write the environmental policy. Although there may be no argument with such an approach, top management's job in this case is to have the policy written according to the intentions and principles delineated by the leadership. Regardless of who actually drafts the document, it must be signed by top management; this attests that the stated terms are the policy of the organization and that top management will ensure that the policy is implemented and carried out.

This document should flow into, and be in harmony with, the organization's vision statement and guiding principles (assuming the organization has developed these as part of its strategic plan). Consequently, it will be tailored to the organization's unique business and environmental situation.

---

### ISO 14000 INFO

*"An environmental policy establishes an overall sense of direction and sets the principles of action for an organization. It sets the goal as to the level of environmental responsibility and performance required of the organization, against which all subsequent actions will be judged."*
    *ISO 14004, Clause 4.1.4*

In an organization's environmental policy, ISO expects to see that the organization has clearly stated what it intends to do and what its guiding principles are relative to the environment. This is similar to the *vision statement* and *guiding principles* in Total Quality Management. The intentions set the vision for environmental performance; they are performance ideals. The principles establish the ground rules, or the environmental code of conduct, for all employees; they are behavioral expectations. The intentions and principles guide action and provide the framework from which **environmental objectives and targets** "flow down," or develop and are communicated to and involve employees at all levels. The environmental policy will be the organization's top-level EMS document; all other documents must support it.

ISO 14004, Clause 4.1.4, lists the following items to be considered in an environmental policy:[2]

- the organization's mission, vision, core values, and beliefs
- requirements of and communication with **interested parties**
- continual improvement
- **prevention** of pollution
- guiding principles
- coordination with other organizational policies (e.g., quality, occupational health and safety)
- specific local or regional conditions
- compliance with relevant environmental regulations, laws, and other criteria to which the organization subscribes

---

## ISO 14001, Clause 4.2 a)

[Management must ensure that the environmental policy] is appropriate to the nature, scale and environmental impacts of its activities, products or services. . . .

---

Requirement—Clause 4.2 a)

The environmental policy must accommodate the organization's business and environmental position. This means, among other things, that the organization has to understand its current propensity for environmental impact in all its activities, products, and services. Furthermore, it must develop an environmental policy that is compatible with this propensity for environmental impact and the nature and scale of the business.

Application Information—Clause 4.2 a)

In order to establish the organization's current environmental position, ISO 14004, Clause 4.1.3, suggests an *initial environmental review* covering the following:[3]

- Identification of legislative and regulatory requirements. Which environmental laws and regulations apply to the organization's operations?

- Identification of environmental aspects of its activities, products, or services in order to determine those that have or can have significant environmental impacts and liabilities. Chapter 2 explained that an environmental aspect is anything resulting from an organization's activities, products, or services that may result in an environmental impact. Further clarification may be found in the text related to Clause 4.3.1 and in ISO 14004, Clause 4.2.2.

- Evaluation of performance compared with relevant internal criteria, external standards, regulations, codes of practice, and sets of principles and guidelines. How does the organization's current environmental performance compare with the various internal and external performance requirements and criteria?

- Existing environmental management practices and procedures. What environmental **practices and procedures** does the organization currently use?

- Identification of existing policies and procedures dealing with procurement and contracting activities. The intent is that the organization's environmental criteria be flowed down to the organization's suppliers.

- Feedback from the investigation of previous incidents of noncompliance. If the organization has experienced legal or regulatory environmental noncompliance in the past, it should learn from it. How did it happen? What was done to prevent recurrence? Is there a chance of future noncompliance?

- Opportunities for competitive advantage. What is the organization doing now that if changed could result in a **competitive advantage** through reduced cost, shorter cycles, improved product or service, decreased liability, or other factors that enhance competitiveness?

- Views of interested parties. What do the organization's employees, customers, suppliers, neighbors, and other affected parties have to say about its current environmental performance?

- Functions or activities of other organizational systems that can enable or impede environmental performance. What are the organization's internal enablers and inhibitors of satisfactory environmental performance? Are there any external enablers or inhibitors?

Having this information will enable the organization to establish a baseline for its current environmental position and realistically develop its environmental policy.

The next step is to look at the organization's business. If the organization were an accounting firm, it would be ridiculous to concern itself with air or water pollution. There might be, however, some environmental considerations for office situations; for example, disposal of toner cartridges, depletion of resources, landfill impacts, and others. In developing its environmental policy, the organization should consider its financial posture. A firm that is barely surviving would be ill advised to indulge in very expensive cutting-edge technology to eliminate minor traces of pollution. In other words, address the relevant issues and be realistic.

## ISO 14001, Clause 4.2 b)

[Management must ensure that the environmental policy] includes a commitment to continual improvement and prevention of pollution. . . .

### Requirements—Clause 4.2 b)

The first two of three things that *must* be included in the organization's environmental policy document are (1) *commitment* to continual improvement and (2) *commitment* to the prevention of pollution.

### Application Information—Clause 4.2 b)

ISO is to be commended for embracing **continual improvement** in ISO 14000. Total Quality Management practitioners have been disappointed with ISO through two succeeding versions of the ISO 9000 Quality Standard for not promoting the concept of continual improvement. Continual (or *continuous*) improvement is one of the essential elements of Total Quality Management and is based on the Shewhart Cycle, also called the Deming Cycle or the Plan-Do-Check-Act (PDCA) Cycle.[4] Figure 3-1 represents the traditional configuration of the **PDCA Cycle**. ISO 14001, Annex A, Section A.1, states:

> This International Standard contains management system requirements, based on the dynamic cyclical process of "plan, implement, check and review."

*Implement* and *review* are substituted for *do* and *act*, respectively. (The authors also routinely substitute *adjust* for *act*.) Regardless of terms used, they refer to the venerable Shewhart/Deming/PDCA Cycle. Unfortunately, neither Shewhart nor Deming is credited in the ISO documents with the cycle's invention. The PDCA Cycle operates like this:

1. **Plan** the improvement.
2. **Do** (or implement) the planned improvement.

**Figure 3-1**
The PDCA Cycle

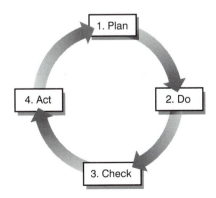

3. **Check** the results to determine whether the anticipated improvement occurred.

4. Based on the results, **adjust** the improvement by going through the cycle again. If it is already satisfactory, make the improvement permanent and monitor it.

5. Continue the cycle, refining the improvement, or developing new improvements.

The model for ISO's **environmental management system** is a variation of the PDCA Cycle that puts it into an organizational context, as illustrated in Figure 3-2. Compare this model with the one in Figure 3-1.

Through continual improvement of the EMS, the processes and procedures involved, the training of affected personnel, the equipment and technology employed, and all other relevant factors, the organization will improve both the efficiency and effectiveness of its environmental efforts. In other words, just as quality improvements lead to a better competitive position, so will improvements in environmental performance.

In Clause 4.2 b) the second requirement for the organization's environmental policy is commitment to the prevention of pollution. Not only must this commitment be stated in the environmental policy, it must be a visible, sincere part of management's activity in order to be effective. If such a commitment by top management is not apparent to all employees, the employees might apply only lip service to environmental efforts. One of the authors' favorite management maxims is:

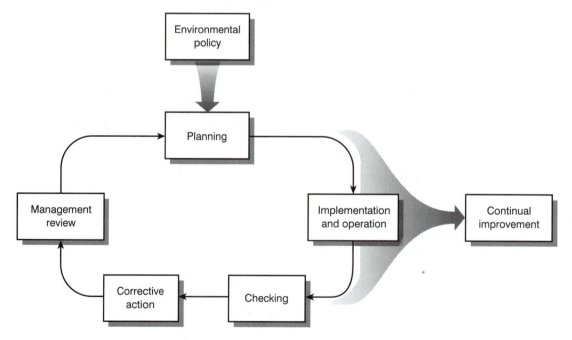

**Figure 3-2**
ISO 14000 Environmental Management System Model

Adapted from ISO 14001, Figure 1

*Employees tend to do what they believe is important to management.*

Employees observe what management does and how management spends its time. If the employees do not see management personally and actively involved in an effort, they conclude that the effort must be unimportant and, consequently, are unlikely to put much effort into it.

Without top management's commitment, the organization's environmental efforts will fail. Although midlevel managers may try, only top management can bring about the necessary alignment of activities or commit funds and human resources necessary to develop and operate a viable EMS.

---

## ISO 14001, Clause 4.2 c)

[Management must ensure that the environmental policy] includes a commitment to comply with relevant environmental legislation and regulations, and with other requirements to which the organization subscribes. . . .

---

### Requirements—Clause 4.2 c)

This clause states the third requirement for a top-level commitment in the organization's environmental policy. It must contain a statement of management's commitment to comply with all relevant environmental legislation, regulations, and any other requirements applicable to the organization, industry, and locale.

### Application Information—Clause 4.2 c)

Clause 4.2 c) can be easily understood, except in its final phrase. "Other requirements" may include items such as permits, licenses, health and safety regulations, and any requirements beyond those legally imposed to which the organization subscribes. Some of these may be voluntary programs. Use caution when considering compliance to voluntary programs. The registrar's auditors will hold the organization accountable to these voluntary programs once the organization commits to comply with them, just as if the voluntary programs were legal requirements.

Management is legally bound to observe all applicable legal and regulatory requirements, so gaining a commitment to comply should not be difficult. In reality, though, scarcely a week goes by without reports of another organization in trouble for gross non-compliance. The bigger problem, especially for small businesses, may be identifying all applicable laws and regulations. This is best achieved by talking to representatives of local, county, state, and federal agencies. Industry and trade associations can also assist in identifying all the legislative and regulatory requirements affecting an organization.

---

## ISO 14001, Clause 4.2 d)

[Management must ensure that the environmental policy] provides the framework for setting and reviewing environmental objectives and targets. . . .

---

### Requirements—Clause 4.2 d)

An organization must establish environmental objectives and targets. These should aim at compliance with legislative and regulatory requirements and be based on the philosophy and intentions committed to in the environmental policy. The clause requires that the policy be developed in a way that facilitates a downward flow of environmental objectives and targets, the strategic and tactical goals, and that the objectives and targets support the policy. The policy should also provide for periodic progress reviews in reference to meeting its objectives and targets.

### Application Information—Clause 4.2 d)

The following example clarifies how a policy should be developed in order to facilitate a downward flow of objectives and targets. The production processes of Organization ABC generate a lot of heat in certain machines. Therefore, one of its *activities* is cooling the machines by drawing water from an adjacent stream and pumping it through the water jackets of the machines. Once the cooling cycle is completed, during which the water becomes hot, it is discharged back into the stream. The *cooling activity* was identified as having an *environmental aspect*: the heating and discharging of the cooling water. Organization ABC recognizes that this *aspect* could cause a harmful *impact* on the stream's ecology. Consequently, the organization wrote into its new environmental policy that it was "committed to minimize any ecological impact to the environment." That statement and others like it provide the framework for developing and tracking objectives and targets.

Once Organization ABC made several commitments in its environmental policy, it had to ask how the intentions stated in the policy might be achieved. The answer to such questions are **objectives**. For example, in reference to the equipment cooling activity, environmental experts concluded that if the temperature of the discharged water could be maintained in the range of 7–16°C, impact to the stream would be minimal. Management accordingly set an *objective* for controlling the temperature of discharged water. In effect, this was a strategy for supporting the organization's environmental policy and commitment to minimize ecological impacts (i.e., by controlling discharged water temperature).

The organization had an objective that had "flowed down" from the environmental policy. At that point, though, specifically *what* had to be done to control the temperature of discharged water was an open question, and several possibilities existed. Organization ABC adopted the solution to circulate the water through a cooling tower before discharging it into the stream. This approach offered potential for achieving the necessary control without the need to purchase a high tech system that the organization could not afford. The organization set the *target* for having the cooling tower system on-line by a specified date. This target, with its explicit terms, represented a tactical objective (i.e., "This is *what we will do* to achieve the strategic objective of controlling discharged water temperature."). This example is exactly what ISO 14001, Clause 4.2 d), requires: environmental policy statements, intentions, principles, and philosophy that lead to, or provide the framework for, setting objectives and targets.

Setting objectives and targets completes the first half of the Clause 4.2 d) requirement. As the second half, objectives that are set must be achieved. This can require considerable time and effort. It is critical that management keep itself apprised of progress. ISO provides no guidance on this, but at least a monthly review of all objectives is appropriate, and the objectives should be evaluated in the context of the environmental policy statements. In doing this, management stays informed about actual performance measured against the objectives and the policy. Also important, employees can see that actual performance is important to management.

---

## ISO 14001, Clause 4.2 e)

[Management must ensure that the environmental policy] is documented, implemented and maintained and communicated to all employees. . . .

---

### Requirements—Clause 4.2 e)

Clause 4.2 e) contains four explicit requirements:

1. The environmental policy must be documented
2. The environmental policy must be implemented
3. The environmental policy must be maintained
4. The environmental policy must be communicated to all employees

### Application Information—Clause 4.2 e)

*Documenting* the environmental policy means that it must be in writing as a formal document, either in paper or electronic format, as long as it is readily available to anyone who needs it. Most organizations use paper documentation; however, it is common practice to keep the documents in electronic format and distribute hard copies to the various departments as required.

*Implementing* the environmental policy means that once the environmental policy is defined, it is put into practice. Management must provide the leadership, but *everyone* who can in any way affect the organization's environmental performance must be involved.

ISO states that an environmental policy must be *maintained*, or be active. Management must declare to everyone that the environmental policy is not to be ignored. It is a living, ongoing policy through which the organization conducts its environmental life; it is always in effect. This is another case where leadership is essential.

The environmental policy must be *communicated*. In addition to the challenges associ-ated with effective **communication**, some managers do not understand how important it is that all employees, not just a select few, know and understand policies and procedures. It is understandable that employees find it difficult to follow a policy or procedure about which they know nothing. It is, likewise, always a challenge to communicate information to all employees, and then to ensure that they absorb and understand it. However, management must apply work, time, the use of every type of media available, and especially repetition to communicate effectively. These things are never-ending management responsibilities.

---

## ISO 14001, Clause 4.2 f)

[Management must ensure that the environmental policy] is available to the public.

---

### Requirements—Clause 4.2 f)

Once an organization defines and documents its environmental policy, the policy must be made available to the public.

### Application Information—Clause 4.2 f)

Managers might consider it unusual to make an internal policy available outside the organization; in fact, seldom do private sector organizations have to make their policies available to the public. Where the environment is concerned, however, every person who has an interest in the air, land, or water in or near the organization is a stakeholder. The community in which the organization operates clearly has the right to know where it stands relative to pollution and the environment, as do the legal and regulatory agencies, customers, suppliers, employees, stockholders, and all other stakeholders.

ISO does not specify how the environmental policy is to be made available to the public. ISO does not expect copies to be distributed to every household, business, or organization in an area; rather, copies can be made available for pickup at a reception desk or at some accessible location on the organization's premises. As an alternative, copies for public review might be distributed to local libraries, chambers of commerce, or other public access organizations. The key is to ensure that if someone asks to see the environmental policy, a copy is made available readily.

### Summary of ISO 14001, Clause 4.2

ISO 14001, Clause 4.2, and Subclauses a) through f) state the environmental manage-ment system requirements for commitment and policy elements. They are:

- Top management is responsible for defining the organization's environmental policy.
- The environmental policy must be appropriate for the business.
- The environmental policy must include a commitment to continual improvement and prevention of pollution.

- The environmental policy must include a commitment to comply with relevant legislation, regulations, and other requirements.

- The environmental policy must provide the framework for setting and reviewing environmental objectives and targets.

- The environmental policy must be documented, implemented, maintained, and communicated to all employees.

- The environmental policy must be available to the public.

## Requirements Section 3: Planning

The second EMS principle states, "An organization should formulate a plan to fulfill its environmental policy." Once the environmental policy is developed, the organization has a clear picture of its intentions for environmental performance. The question remains, however: How do we make it happen? It is not sufficient for the environmental policy to commit the organization to compliance with the law and regulations, relying on the intelligence, skill, and judgment of all employees to do the right thing every time. There has to be an environmental plan that can be implemented, followed by the employees, tracked by management, and audited by interested parties. Planning should look at each of the elements of the EMS. ISO 14004, Clause 4.2.1, lists the following as relating to planning:

- Identification of **environmental aspects** and evaluation of associated environmental impacts
- Legal requirements
- Environmental policy
- Internal performance criteria
- Environmental objectives and targets
- Environmental plans and management program

The paragraphs that follow illustrate that ISO 14001, Clause 4.3, Planning, addresses each of these EMS elements directly, except for environmental policy, already covered by Clause 4.2, and internal performance criteria, which is covered under "other requirements."

---

## ISO 14001, Clause 4.3: Planning

### ISO 14001, Clause 4.3.1: Environmental Aspects

The organization shall establish and maintain (a) procedure(s) to identify the environmental aspects of its activities, products or services that it can control and over which it can be expected to have an influence, in order to determine those which have or can have significant impacts to the environment. The organization shall ensure that the aspects related to these significant impacts are considered in setting its environmental objectives. The organization shall keep this information up to date.

### Requirements—Clause 4.3.1

This clause requires three things. First, it issues a mandatory requirement (*shall*) to establish and employ procedures for identifying its environmental aspects and significant impacts. Second, it stipulates that the organization must ensure that the aspects related to significant impacts are considered in establishing environmental objectives. Third, it defines that the organization must keep its aspect and impact information up-to-date.

### Application Information—Clause 4.3.1 Requirements

In regard to the first requirement, the clause states that "The organization shall establish and maintain (a) procedure(s) to identify the environmental aspects of its activities, products or services . . . in order to determine those which have or can have significant impacts to the environment." *Shall* means that this is an explicit requirement—it must be done. Further, "establish . . . (a) procedure(s)" means the organization must develop or create a procedure, or procedures if more than one is required. The statement "maintain (a) procedure(s)" means, as before, maintaining the use of the procedure(s). It also allows for revising the procedure(s) as practice and use would indicate.

It is not required that the procedure(s) be documented. In very small organizations documented procedure(s) may not be appropriate. However, it is difficult to think of an instance where having the procedure(s) documented would not be preferred. For practical purposes the procedure(s) should be documented.

ISO is not limiting the requirement for identification of aspects and impacts to those associated with the organization's *activities*. The organization also must identify those associated with its *products* and/or *services*. One could easily imagine scenarios that would pose problems for the organization. For example, if the organization is a manufacturer of automobiles, it is one thing to develop procedures to be used in the plant for preventing pollution during the manufacturing processes. It is quite another problem for it to deal with the potential environmental impact of the product once it is delivered to its end user. All that an automobile manufacturer can realistically do is ensure that its in-plant processes are the cleanest possible, its cars comply with existing emission regulations, and its customers are advised to follow prudent maintenance procedures. Organizations are not responsible for what their customers might or might not do. What relieves them of responsibility? Notice the phrase in the clause that says, " . . . that it can control and over which it can be expected to have an influence. . . . " Neither ISO nor any legislative body or regulatory agency known holds the manufacturer of a product responsible for a customer's actions in terms of environmental impact. Interestingly though, we cannot say the same for safety issues.

ISO 14004 lists nine issues that should be considered in an organization's procedure(s) for identification of environmental aspects and evaluation of environmental impacts.[5]

1. What are the environmental aspects of the organization's activities, products, and services?

2. Do the organization's activities, products, or services create any significant adverse environmental impacts?

3. Does the organization have a procedure for evaluating the environmental impacts of new projects?

4. Does the location of the organization require special environmental consideration, for example, sensitive environmental areas?

5. How will any intended changes or additions to activities, products, or services affect the environmental aspects and their associated impacts?

6. How significant or severe are the potential environmental impacts be if a process failure occurs?

7. How frequently will the situation arise that could lead to the impact?

8. What are the significant environmental aspects, considering impacts, likelihood, severity, and frequency?

9. Are the significant environmental impacts local, regional, or global in scope?

In Chapter 2 we defined an environmental aspect as any element of an organization's activities that could have an impact on the environment, beneficial or detrimental. An environmental impact is a change that takes place in the environment as a result of an aspect. For example, of the many activities taking place in an auto repair shop, several of them have environmental aspects. One of the shop's activities is oil changes. The *activity* of changing oil brings with it the potential for the *aspect* of spilling oil. The aspect of spilling oil carries the potential for *environmental impact* by contaminating the soil or water. Notice the cause-and-effect relationship between the aspect and the impact; the aspect is the cause, and the impact is the effect. To satisfy this clause of ISO 14001, the organization must establish a procedure, then use it, to identify its environmental aspects and any significant potential environmental impacts. The general procedure to do that is as follows:[6]

1. Select an activity, product, or service to examine. Keep the selection narrow enough to be understood and manageable. For example, if the organization is a manufacturer, selecting "manufacturing" would be too broad; it would be appropriate to select a manufacturing process, such as grinding or plating. The same kind of thinking should be used for products or services. Hotel maintenance would be too broad; instead, break down hotel maintenance into its constituent parts.

2. Next, examine the selected activity, product, or service for as many environmental aspects as possible. Again, the aspect is anything resulting from the activity (or product or service) that can have an impact on the environment.

3. For each environmental aspect listed, identify all possible (actual and potential) environmental impacts, both positive and negative.

4. Evaluate each impact for severity, the probability that it will occur, and its probable duration. Also evaluate it from the business standpoint. That is, if a particular impact should occur, determine whether there would be legal or regulatory exposure, how difficult and costly it would be to change the impact, what the likely concerns of interested parties would be, and how public image would be affected.

ISO expects an organization that does not already have an EMS to establish its current environmental position through a review covering four key areas:[7]

1. Legislative and regulatory requirements
2. Identification of significant environmental aspects
3. Examination of all existing environmental management practices and procedures
4. Evaluation of feedback from the investigation of previous incidents

A review should consider abnormal and emergency conditions, shutdown and start-up, and normal operations. The identification of significant aspects is not expected to include a detailed evaluation of all possible activities, products, or services, but only those which are considered most likely to have significant environmental impact.

In regard to the second requirement, the organization must (*shall*) consider all aspects that could have significant impacts in establishing its environmental objectives. If an aspect is listed as significant, the organization should have a corresponding environmental objective. The objective would be to eliminate the aspect from the significant list or find a way to ensure that the impact never occurs. ISO 14001 does not explicitly require an objective for each significant aspect, only that significant aspects be considered when setting objectives. If for some reason an objective is not set for a significant aspect, the prudent organization will be prepared to show auditors that it was considered.

In regard to Clause 4.3.1's third requirement, the organization must (*shall*) maintain up-to-date information about its environmental aspects and impacts. Although there is no explicit requirement for the information to be documented, as a practical matter there is no other way to keep it up-to-date. It should be either in electronic format or on paper, or both.

---

## ISO 14001, Clause 4.3.2: Legal and Other Requirements

The organization shall establish and maintain a procedure to identify and have access to legal and other requirements to which the organization subscribes, that are applicable to the environmental aspects of its activities, products or services.

---

Requirements—Clause 4.3.2

The organization must develop and continually use a procedure that identifies and provides access to:

1. Environmental legal (i.e., legislative and regulatory) requirements.
2. All other environmental requirements to which the organization has agreed to comply.

There is an implicit requirement that the organization must "understand" applicable legal and other requirements.

## Application Information—Clause 4.3.2

The organization must have a procedure. It can be developed in-house or by an outside consultant. The requirement is that the organization have and use a procedure which will ensure that all applicable legal and other environmental requirements are identified, updated as required, and accessible (usually in the form of documents) to people within the organization who need them. It is necessary to identify and put into its environmental library only those legal and other requirements that are applicable to the significant environmental aspects of the organization's activities, products, and services—not *every* environmental statute, code, and practice.

The second requirement of this clause is that the procedure be maintained to ensure that legal and regulatory changes are noted and reflected in the organization's activities. Such maintenance also will ensure that any changes in the organization's operations affecting its significant environmental aspects will result in the identification of any relevant new legal or other requirements.

While there is no explicit requirement, there is certainly an implicit requirement that a legal or other environmental requirement, once identified, be understood by the organization. ISO expects organizations to identify and comply with legal and other requirements, and to do so a thorough working knowledge of them is essential. ISO wants assurance that the organization will identify and maintain copies of the environmental requirements and that everyone in the organization who could possibly have an impact understands the requirements.

"Legal" environmental requirements are generally understood to be those coming from governmental legislative and regulatory action. "Other" requirements can include industry codes of practice, contracts, agreements with public authorities, and nonregulatory guidelines.[8] Some of these may be binding; some may be voluntary. However, if the organization has agreed to abide by them in its environmental policy, then ISO requires that they be treated like the legal requirements. To reiterate, the auditors will hold the organization accountable for what it said it would do.

Issues to be considered in an organization's procedure under this clause should include how the organization:[9]

1. accesses and identifies legal and other requirements,
2. keeps track of legal and other requirements,
3. keeps track of changes to legal and other requirements, and
4. communicates relevant information about legal and other requirements to employees.

Regulations exist in many forms. There are environmental laws that apply to everyone, and there are those specific to an industry (i.e., the shipping industry relative to purging of fuel tanks and disposing of garbage). Others apply to particular products or services, such as emission-control systems on motor vehicles. Some authorize a business, or a site or activity of a business, through licenses or permits. Organizations, particularly smaller ones that do not have legal staffs, can have difficulty in identifying all legal and other environmental requirements that apply to the environmentally significant aspects of their activities. Resources that can help include:

- Local, state, and federal governments
- Industry and trade associations
- Consulting services
- Outside databases

---

## ISO 14001, Clause 4.3.3: Objectives and Targets

The organization shall establish and maintain documented environmental objectives and targets, at each relevant function and level within the organization.

When establishing and reviewing its objectives, an organization shall consider the legal and other requirements, its significant environmental aspects, its technological options and its financial, operational and business requirements, and the views of interested parties.

The objectives and targets shall be consistent with the environmental policy, including the commitment to prevention of pollution.

---

### Requirements—Clause 4.3.3

This clause establishes mandatory requirements and sets some procedural requirements for doing so. The mandatory requirements are:

1. The organization must develop environmental objectives and targets.
2. Objectives and targets must be documented.
3. Objectives and targets must be developed for each relevant function and level within the organization.
4. Established objectives and targets must be acted upon and reviewed.

The procedural requirements are:

1. Objectives and targets must be consistent and compatible with the environmental policy, significant environmental aspects, and legal and other environmental requirements.

---

### ISO 14000 INFO

Objectives *are established to meet the organization's environmental policy. They are the overall goals for environmental performance identified in the environmental policy.* Targets *are set to achieve these objectives within a specified time frame and should be specific and measurable.*[10]

2. Technological options, the organization's financial, operational, and business requirements, and the views of interested parties must be considered.

## Application Information—Clause 4.3.3

Thus far the organization has established three elements upon which the environmental management system is based: its environmental policy, environmental aspects, and legal and other environmental requirements. These three elements provide the framework for establishing environmental objectives and targets, which provide the road map for everything the EMS accomplishes. The policy and significant aspects "flow down" into objectives and targets. The targets specify what must be done to achieve the objectives. As the objectives are accomplished, the organization moves closer to realizing the policy. This results in continual improvement of environmental performance.

Therefore, the organization's *first* requirement under this clause is to look at its legal and other requirements, significant environmental aspects, and its environmental policy and then develop a set of objectives and targets. Every environmental policy must have a commitment to prevent pollution. To illustrate, an organization might have an aspect that could impact the environment by emitting a harmful gas. A corresponding objective might require the elimination of the possibility of emitting that gas. A corresponding target could propose the redesign of the process in order to eliminate use of the gas. This kind of solution (i.e., eliminating the use of a polluting gas) also eliminates the risk of accidental release of the gas, and thereby reduces potential liabilities—always a good thing.

The objective may be broad or specific in scope, and it does not have to be directly measurable. The target specifies what to do to satisfy the objective, and it will almost always be measurable in terms of change, elimination, or date. The objectives and targets relate to the organization's stated intentions and philosophy in the environmental policy as well as its significant environmental aspects.

In addition to the legal and other requirements and significant environmental aspects, this clause also requires that the organization consider:

- The best technological option to mitigate an aspect.
- Economic viability of the option.
- Cost-effectiveness of the option.
- Appropriateness of the option to the situation.
- Affordability of the option, given the organization's financial, operational, and business situation.
- Views of interested parties.

The final point here is important when the organization's activities and environmental aspects are scrutinized by the community. ISO requires that the input of stakeholders be considered in the establishment of environmental objectives and targets. Interested parties can include legislative and regulatory agencies, employees, and anyone with a legitimate stake in the issue.

The *second* requirement of this clause is that the organization's objectives and targets be documented as formal documents, either on paper or in electronic format.

The clause's *third* requirement is that each function (i.e., department) of the organization which has any influence on environmental performance must have environmental objectives and targets. This could mean every department. Even if a department, say accounting, does not see itself as directly contributing to pollution, it probably does; and it certainly consumes resources, which are an ISO 14000 consideration. The question of whether these are *significant* aspects of the organization's activities, products, and services will have to be answered by the organization.

The *fourth* requirement states that the organization must periodically review its objectives and targets. Management has to establish some sort of review program with persons responsible for implementing the objectives and targets. This review is an important part of the EMS model based on the PDCA Cycle discussed earlier. The reviews will let management know if the objectives and targets are on schedule and producing the desired results. Management will then be able to make decisions to continue the implementation of specific objectives and targets or to introduce needed changes.

Objectives may apply broadly across the entire organization or narrowly to one function or even one person in the organization. They are best established with participation of operating levels that deal daily with the situation being addressed. Considerations to be applied to environmental objectives and targets include:

1. The organization's environmental policy and significant environmental aspects and impacts should be reflected.
2. Employees responsible for achieving the objectives and targets should participate in their development.
3. Views of interested parties should be heard and considered.
4. Objectives and targets should have measurable indicators, if possible.
5. Regular management reviews should occur to track progress and evaluate performance improvements resulting from objective achievement.

Objectives can include many forms of commitments. ISO 14004, for example, lists:[11]

- Reduce waste and the depletion of resources
- Reduce or eliminate the release of pollutants into the environment
- Design products to minimize their environmental impact in production, use, and disposal
- Control the environmental impact of sources of raw material
- Minimize any significant adverse environmental impact of new developments
- Promote environmental awareness among employees and the community

The final statement of Clause 4.3.3 states that objectives and targets must be consistent with the environmental policy, including the commitment to prevent pollution.

Does this mean that every objective or target must prevent pollution? One could certainly interpret it that way. Joseph Cascio, chairman of the U.S. Technical Advisory Group to ISO/TC 207, says that the organization has to show only that it considered prevention of pollution in its activities, products, and services. Although objectives and targets might show evidence of pollution prevention, it is not a requirement.[12]

---

## ISO 14001, Clause 4.3.4: Environmental Management Program(s)

The organization shall establish and maintain (a) program(s) for achieving its objectives and targets. It shall include

    a) designation of responsibility for achieving objectives and targets at each relevant function and level of the organization

    b) the means and time-frame by which they are to be achieved.

If a project relates to new developments and new or modified activities, products or services, program(s) shall be amended where relevant to ensure that environmental management applies to such projects.

---

### Requirements—Clause 4.3.4

Clause 4.3.4 states two requirements.

1. The organization *must* create and implement one or more programs designed to assure that its objectives and targets are achieved.
2. To achieve the objectives/targets, the program(s) *must* provide for:
   - assignment of responsibility to the relevant function or individual
   - logistics, methods, and resources
   - schedule
   - applicability to new developments and new or modified activities, products, or services

### Application Information—Clause 4.3.4

Some confusion is associated with this clause. Does the clause require a separate program for each objective and target, or one program designed to deal with all the organization's environmental objectives and targets? The clause itself uses the phrase, "(a) program(s)." Surely no organization would be expected to be limited to one objective or target. Therefore, one might interpret the requirement to be *either* "one program for all objectives," or "a separate program for each objective." Joseph Cascio et al. in *The ISO*

*14000 Handbook* states that this clause addresses the " . . . individual program(s) that need to be established and implemented to achieve your objectives and targets," and says, "You need a program for each stated objective."[13] This appears to mean a *separate* program for each objective and target. On the other hand, ISO 14004, Clause A.3.4, provides this insight:

> The creation and use of *one or more programs* is a key element to the successful implementation of an environmental management system. *The program* should describe how the organization's *objectives and targets* will be achieved,. . . . " (emphasis added)

In this case a single program clearly addresses multiple objectives and targets. Either approach can be used, and either should satisfy the registrar. Since registrars may interpret this differently, it would be prudent for the organization to determine how its registrar would like to see this clause executed.

This clause intends that a system be devised and implemented to ensure that the objectives and targets established by the organization are realized. ISO does not offer much more guidance than to state certain requirements for inclusion in such a system:

1. Responsibility for objectives and targets must be designated. If more than one department or individual shares the responsibility, then the responsibilities of each must be clarified.
2. The manner in which the objectives and targets are to be achieved must be described.
3. A schedule for achieving the objectives and targets must be included.
4. The system must accommodate new developments and new or modified activities, products, or services as they surface.

This might best be accomplished by an overall **environmental objective and target achievement program**, a name the authors find more descriptive than ISO's *environmental management program*. Whatever name it is called, the program should outline the general scheme by which all environmental objectives and targets are to be achieved. Figure 3-3 shows a synopsis of the process for accomplishing all environmental objectives.

The seven-step process can be an integral element of the EMS becoming the single environmental management program for achieving all of the organization's environ-

---

**ISO 14000 INFO**

*Another point of possible confusion exists with this clause. The reader should understand that the environmental management* program *is not the same thing as the environmental management* system. *The environmental management program becomes an element of the environmental management system. This should be self-evident, but it has been misinterpreted even by a well-informed writer in* ISO 14000: A Guide to the New Environmental Management Standards.[14]

---

**Environmental Objective/Target Achievement Program Process**

Step 1.    State the objective/target.

Step 2.    State the purpose of the objective/target. (How does the objective/target support the environmental policy?)

Step 3.    Describe how the objective/target is to be achieved.

Step 4.    Designate the Objective/Target Program Leader.

Step 5.    Designate departments and individuals responsible for specific milestones.

Step 6.    Establish schedule for milestones and completion of objective/target.

Step 7.    Establish Objective/Target Achievement Program review format, content and schedule.

---

**Figure 3-3**
Steps of the Environmental Objective and Target Achievement Program

mental objectives and targets. However, each objective and target is subjected to the seven steps and becomes a distinct program at the objective or target level, thus meeting the requirements of the clause under either interpretation. The same seven-step process can be used for new or modified activities, products, or services and to incorporate new developments that may be applicable to environmental performance. The Environmental Objective and Target Action Plan form of Figure 3-4 was developed to facilitate and document the environmental objective and target achievement program.

To illustrate the use of this process, an electronics plant has determined that it must eliminate CFC emissions in order to comply with new emission regulations and to conform to its new ISO 14000 environmental policy, which pledges to prevent pollution. The company established the elimination of CFC emissions as an environmental objective. It concluded that the best way to do this was by changing their printed circuit board cleaning process to eliminate the use of CFCs. With the addition of a date, this objective became the target. Then the organization used the seven-step Environmental Objective and Target Achievement Program (Figure 3-3) and the Environmental Objective and Target Action Plan (Figure 3-4) to convert the objective and target into a program. The resulting plan is shown in Figure 3-5.

Following the seven-step process, the **first entry** on the Environmental Objective and Target Action Plan lists the objective and target. This satisfies Clause 4.3.3, Paragraph 1. (In this example the target, a new process, was known at the time the form was initially completed. On some occasions the objective is clear, but the action necessary to achieve the objective, the target, needs to be determined. In that case, the form is filled in where it can be, and one of the tasks for the program leader may be to determine the best tactical approach, which then becomes the target. A single objective may have several targets, in which case separate forms [programs] would be used for each target. In some cases management may designate tasks and schedules, and in others this may be determined by the program leader.)

---

**Environmental Objective/Target Action Plan**

1. Objective:_____

   Target: _____

2. Accomplishing Objective/Target Will: _____

   _____

   _____

3. Plan for Accomplishing: _____

   _____

   _____

4. Program Leader: _____

5. Tasks                          Dept.              Assigned to

   _____      _____      _____

   _____      _____      _____

   _____      _____      _____

   _____      _____      _____

6. Schedule



7. Objective/Target Program Review

   Format: _____

   Required Content: _____

   Review Schedule: _____

                                   Use additional sheets as required.

---

**Figure 3-4**
Environmental Objective and Target Action Plan

The **second entry** of the form documents the purpose of the objective and target and confirms that it supports the environmental policy. This documentation satisfies Clause 4.3.3, Paragraph 3.

The **third entry** provides a brief plan for accomplishing the objective and target. Additional pages may be attached to the form if more space is needed.

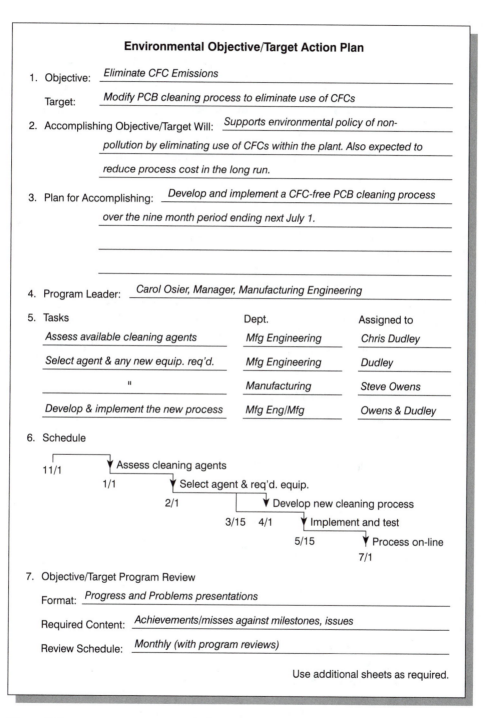

**Environmental Objective/Target Action Plan**

1. Objective: _Eliminate CFC Emissions_

   Target: _Modify PCB cleaning process to eliminate use of CFCs_

2. Accomplishing Objective/Target Will: _Supports environmental policy of non-pollution by eliminating use of CFCs within the plant. Also expected to reduce process cost in the long run._

3. Plan for Accomplishing: _Develop and implement a CFC-free PCB cleaning process over the nine month period ending next July 1._

4. Program Leader: _Carol Osier, Manager, Manufacturing Engineering_

5. Tasks

| Tasks | Dept. | Assigned to |
|---|---|---|
| _Assess available cleaning agents_ | _Mfg Engineering_ | _Chris Dudley_ |
| _Select agent & any new equip. req'd._ | _Mfg Engineering_ | _Dudley_ |
| _"_ | _Manufacturing_ | _Steve Owens_ |
| _Develop & implement the new process_ | _Mfg Eng/Mfg_ | _Owens & Dudley_ |

6. Schedule

   11/1 → Assess cleaning agents
   1/1 → Select agent & req'd. equip.
   2/1 → Develop new cleaning process
   3/15  4/1 → Implement and test
   5/15 → Process on-line
   7/1

7. Objective/Target Program Review

   Format: _Progress and Problems presentations_

   Required Content: _Achievements/misses against milestones, issues_

   Review Schedule: _Monthly (with program reviews)_

   Use additional sheets as required.

**Figure 3-5**
Completed Environmental Objective and Target Action Plan

The **fourth entry** designates the individual responsible for managing the program. This documentation satisfies a requirement of Clause 4.3.4 a).

The **fifth entry** lists specific tasks to be accomplished and assigns them to departments and/or individuals. Continuation sheets may be used if there is insufficient space on the form. This documentation satisfies requirements of Clause 4.3.4 a) and b).

The **sixth entry** is a schedule of key milestones to be accomplished. These will generally conform to the tasks listed in the fifth entry. This documentation satisfies a requirement of Clause 4.3.4 b).

The **seventh entry** establishes the **management review** process for the objective and target achievement program. **Format** refers to the kind of review, such as an oral presentation to management or a written report. In this example the Program Leader will be making a presentation to management. A Progress and Problems presentation is designed to cover progress made (or missed) relative to the plan, problems encountered that may impact the program, and ways in which the team intends to deal with the problems. **Required Content** specifies the information to be covered in the reviews. In this case the reviews must cover milestones met or missed as well as issues that are relevant to objective and target achievement. *Review Schedule* establishes the frequency and timing of required reviews. In our example, management has added the reviews to their monthly business program reviews. Review of these objective and target action plans is necessary to keep management apprised of how well the **environmental management programs** are working, and also to satisfy several ISO 14001 clauses, including 4.2 d), 4.3.3, and 4.6.

Summary of ISO 14001, Clause 4.3 (Planning)

To satisfy all of Clause 4.3 (Planning) the organization must establish and employ procedures to:

- Identify the environmental aspects of its activities, products, or services.
- Identify aspects that can have significant environmental impact.
- Consider aspects with significant impacts in setting environmental objectives.
- Keep aspect and impact information up-to-date.
- Identify legal and other requirements applicable to its environmental aspects.

The organization must also set, act upon, and review environmental objectives and targets, and in so doing:

- Document them.
- Set and deploy objectives and targets for each relevant function (department, section, etc.).
- Consider legal and other requirements.
- Consider its significant environmental aspects.
- Consider its technological options.
- Consider its financial, business, and operational requirements.

■  Consider the views of interested parties inside and out of the organization.

■  Ensure that objectives and targets are consistent with the organization's environmental policy and its commitment to prevention of pollution.

The organization must establish and use structured, planned programs for achieving its environmental objectives and targets. This includes:

■  designation of responsibility at each relevant function and level

■  a plan for how the objective and target are to be achieved

■  a schedule for achievement

New developments and new or modified activities, services, or products must be accommodated by these programs.

## Requirements Section 4: Implementation

ISO 14001, Clause 4.4, with its seven subclauses, is devoted to the third principle of ISO's EMS model, **Implementation**. Implementation is concerned with development of an organizational authority structure, infrastructure, capabilities, and support mechanisms that will enable the organization to achieve its environmental policy, objectives, and targets. In the three requirements sections preceding this one, the organization will have developed its environmental policy, identified its significant environmental aspects and its legal and other environmental requirements, established environmental objectives and targets, and established programs for achieving them. In this section, Implementation and Operation, the organization will address the following seven elements of an EMS:

■  **Organizational structure and responsibilities** (Clause 4.4.1)

■  **Environmental training, awareness, and competence** (Clause 4.4.2)

■  Communication (Clause 4.4.3)

■  **EMS documentation** (Clause 4.4.4)

■  **Control of documentation** (Clause 4.4.5)

■  **Control of operational activities** (Clause 4.4.6)

■  **Emergency preparedness and response** (Clause 4.4.7)

| ISO 14000 INFO |
|---|

## Principle 3—Implementation

*"For effective implementation, an organization should develop the capabilities and support mechanisms necessary to achieve its environmental policy, objectives and targets."*[15]

---

## ISO 14001, Clause 4.4: Implementation and Operation

### ISO 14001, Clause 4.4.1: Structure and Responsibility

Roles, responsibility and authorities shall be defined, documented and communicated in order to facilitate effective environmental management.

Management shall provide resources essential to the implementation and control of the environmental management system. Resources include human resources and specialized skills, technology and financial resources.

The organization's top management shall appoint (a) specific management representative(s) who, irrespective of other responsibilities, shall have defined roles, responsibilities and authority for

- a) ensuring that environmental management system requirements are established, implemented and maintained in accordance with this international Standard;
- b) reporting on the performance of the environmental management system to top management for review and as a basis for improvement of the environmental management system.

---

### Requirements—Clause 4.4.1

Clause 4.4.1 has four primary requirements.

1. The organization must define, document, and communicate the roles, responsibilities, and authority of people at all levels of the organization who may be involved in any way with the EMS.
2. Management must provide the resources (human, skills, technology, and financial) that are required to implement and control the EMS.
3. Top management must designate at least one specific individual, who, irrespective of other responsibilities, will be the EMS **Management Representative**(s).
4. The Management Representative(s) must have defined roles, responsibilities, and authority for (a) ensuring that EMS requirements are established, implemented, and maintained, and (b) reporting on EMS performance to top management for review and as a basis for EMS improvement.

### Application Information—Clause 4.4.1

*Requirement 1*    This requirement intends that the environmental tasks, obligations, and authority to act of everyone involved be clearly defined. Except in very small organizations, there will be many people involved in one way or another with the EMS. Only when all employees understand the tasks, obligations, and authority can the EMS function effectively. ISO wisely used the words *responsibility* and *authority*. It did so to prevent the all-too-common situation where management delegates the responsibility for a task to an employee without assigning him or her the authority necessary to complete it. Under ISO 14000, when management assigns responsibility, the requisite authority must also be given. No hierarchical level is implied in the defining of roles, responsibility, and authority. An effective EMS in most organizations will require the effort of employees at all levels. Thus roles, responsibility, and authority must be defined for all levels of employees who participate in or interact with the EMS.

ISO 14004, Clause 4.3.2.3, states that the following are issues to be considered when assigning responsibilities:

1. What are the responsibilities and accountability of personnel who manage, perform and verify work affecting the environment, and are these defined and documented?
2. What is the relationship between environmental responsibility and individual performance and is this periodically reviewed?
3. How do the responsible and accountable personnel
   - obtain sufficient training, resources and personnel for implementation?
   - initiate action to ensure compliance with environmental policy?
   - anticipate, identify and record an environmental performance?
   - initiate, recommend, or provide solutions to those problems?
   - verify the implementation of such solutions?
   - control further activities until any environmental deficiency or unsatisfactory condition has been corrected?
   - obtain appropriate training to act in emergency situations?
   - gain an understanding of the consequences of non-compliance?
   - gain an understanding of the accountability that applies to them?
   - encourage voluntary action and initiatives?

If management considers these questions in defining roles, responsibilities, and authority for employees at all levels of the organization, the result should be a solid start for the EMS implementation.

Management also must document and communicate the roles, responsibilities, and authority. Again, document means to prepare in written format, whether on paper or in electronic format. The documents may be organization charts, job descriptions, or any other format that will be easily interpreted by all employees. The documents, according to the ISO standard, must be under document control, named, numbered, and dated (to distinguish between the original release and subsequent revisions), and they must be approved by appropriate organizational authority. (Refer to Clause 4.4.5.) In other words, they cannot be notes scribbled on a yellow pad. It would be a good idea to include this documentation in the environmental policy manual, especially for the middle and top levels. Lower levels may be covered in procedure documents.

In addition to defining and documenting the roles, responsibilities, and authority, the information must be communicated. Of course, the employee assigned the role must be informed, but other employees with whom the employee might interact in executing the responsibilities must also know that this person has the responsibility and authority to execute the task. In some cases it is necessary to communicate this information to people out of the organization, for example, fire and police departments. The communication should alert anyone who might need to know that a particular employee is charged with the responsibility and provided with the necessary authority to do the job.

*Requirement 2*    This requirement states that management must provide the resources necessary to implement and control the EMS. While the clause does not specifically state it, before management can make the resources available, the resources required must be identified. Only when management understands quantitatively and qualitatively the resources that will be required can they commit to provide them. Management must determine the EMS's implementation and operational resource requirements in the following categories:

- Human (from which functions, what skills, how many)
- Physical (facilities, equipment, technology)
- Financial

Once management determines the resources required versus resources available, they will be able to make informed decisions about the EMS to which they can commit appropriate resources. If the EMS envisioned would require more resources to implement and control than are available, the EMS may have to be scaled back or reconfigured. The commitment for resources must be made before the EMS implementation is undertaken.

*Requirement 3*    The third requirement states that each registered organization must have an EMS czar. ISO calls the position **Management Representative**, the same as a similar quality management system head under ISO 9000. It does not need to be a full-time assignment; it can supplement other duties. Organizations may appoint more than one Management Representative; this may be appropriate in organizations that have geographically dispersed operations, where one part-time Management Representative would be unable to manage all of the organization's environmental activities.

As far as necessary credentials, ISO 14004, Clause 4.3.2.3, states, " . . . should be assigned to (a) senior person(s) or function(s) with sufficient authority, competence and resources." It may be wise that the Management Representative be the senior manager responsible for environmental activities in the organization. The choice is properly left to management.

*Requirement 4*    The fourth requirement states that management must define the roles, responsibilities, and authority of the Management Representative. These roles, responsibilities and authority are to enable the Management Representative to ensure that EMS requirements are established, implemented, and maintained (continuously met). ISO clearly envisions the Management Representative as the manager who will lead the EMS development, implementation, and operation. We have the same hesitancy with this that we have with the similar ISO 9000 Management Representative, namely, that the introduction of a Management Representative must not be seen as a vehicle for top management to avoid involvement, or to be less involved with the EMS. Top management must remain actively involved for the environmental management system to operate at best efficiency.

If the Management Representative's first order of business is "establishing, implementing, and maintaining" EMS requirements, the second order of business is reporting on EMS performance to top management. Once the EMS is in place and operating, the Management Representative's emphasis will turn to the task of establishing and monitoring environmental objectives and targets through the so-called environmental management programs. It is also important to report on results of the objective initiatives and EMS performance; this feedback fuels management's PDCA Cycle for improvement of the EMS. The Management Representative will in most cases be the organization's point of contact for the registrar, although this is not mentioned in ISO 14001. The Management Representative will be expected to perform duties such as coordinating registrar audits and providing the registrar with information on action items. At a closer look, the Management Representative assignment resembles a full-time job.

---

## ISO 14001, Clause 4.4.2:
## Training, Awareness, and Competence

The organization shall identify training needs. It shall require that all personnel whose work may create a significant impact upon the environment, have received appropriate training.

It shall establish and maintain procedures to make its employees or members at each relevant function and level aware of

a) the importance of conformance with the environmental policy and procedures and with the requirements of the environmental management system;

b) the significant environmental impacts, actual or potential, of their work activities and the environmental benefits of improved personal performance;

c) their roles and responsibilities in achieving conformance with the environmental policy and procedures and with the requirements of the environmental management system, including emergency preparedness and response requirements;

d) the potential consequences of departure from specified operating procedures.

Personnel performing the tasks which can cause significant environmental impacts shall be competent on the basis of appropriate education, training and/or experience.

---

**Requirements—Clause 4.4.2**    The four primary requirements of this clause are:

1. The organization must identify training needs.
2. The organization must require that appropriate training be given to all employees whose work could create a significant environmental impact.

3. The organization must establish and continuously use procedures in order to make each relevant function (job, task) and level aware of
   - the importance of conformance with environmental documentation, including the policy, procedures, and EMS requirements;
   - the significant environmental impacts of its activities (actual or potential) and the environmental benefits of "improved personal performance";
   - the roles and responsibilities concerning achieving conformance with the environmental policy, procedures, EMS requirements, and emergency preparedness and response requirements;
   - the potential consequences of failing to adhere to appropriate procedures.
4. The organization must ensure that it uses only employees who are competent to perform tasks which can cause significant environmental impacts. Competence is based on appropriate education, training, and/or experience.

### Application Information—Clause 4.4.2

Clause 4.4.2 is concerned with two categories of training: competence training and awareness training. All employees whose work could result in a significant environmental impact must receive appropriate skills training where needed. Other employees engaged in less critical environmental functions must be given training to raise their general environmental awareness.

The first requirement of Clause 4.4.2 (note the word *shall*) is for the organization to identify its training needs. These will vary greatly from organization to organization, and will depend largely on hiring practices used, preexisting training policies, processes in use, and employee experience level. Training needs can be determined as follows:

1. Identify the knowledge and skills necessary for each task associated with activities of significant environmental impact potential.
2. Determine the knowledge and skill level of employees who perform those tasks, then identify the gaps.
3. Identify the level of environmental awareness needed by employees.
4. Develop a training plan to address the gaps and awareness training needs.

Identifying training needs should not be considered a one-time activity. After the initial training needs assessment, the organization should repeat it periodically and adjust its training programs accordingly. This is necessary because employees come and go and technology, processes, and environmental requirements change.

The second requirement is that the organization require anyone whose work may create a significant environmental impact to receive appropriate training. This type of training will be the area of skills and knowledge. Employees who have been doing the job acceptably for several years will probably not need much training. A new person, however, or an employee transferred to the job will need more extensive training. Any employee assigned the critical tasks, those with actual or potential significant environmental impacts, should have sufficient training to mitigate the risk of an impact.

The third requirement is that the organization establish and use (maintain) a procedure to develop awareness in employees of certain environmental issues. Therefore, actual awareness training of employees (or members) at each relevant (environmentally related) function and level in the four areas listed in Subclauses a) through d) is required.

In addition to legal and regulatory requirements to which the organization must comply, it must conform to its own environmental policy, operational procedures, and requirements of its EMS. Subclause a) requires that employees associated in any way with environmental activities receive training on the importance of conformance with these internal organizational requirements. Without this awareness training, employees may believe that since the environmental policy, procedures, and the EMS are internally developed, *conformance* to them is not as important as **compliance to legal and regulatory requirements** imposed by the government. They have to understand that the two, **compliance** with legal and regulatory requirements and **conformance** with the internal environmental requirements, go hand-in-hand, and that failure to *conform* to the environmental policy, procedures, or the EMS requirements can result in failure to *comply* with a legal or regulatory requirement.

Subclause b) requires that training be provided to make appropriate employees aware of the significant impacts associated with their work and that it is expected of them to scrupulously adhere to procedures and EMS requirements in order to prevent environmental impact events. These employees must be aware, and appreciate the fact, that their job performance is crucial to avoiding environmental impacts. The prompting in Subclause b) that relevant employees be aware of "the environmental benefits of improved personal performance" would have stirred displeasure in quality pioneer Dr. W. Edwards Deming. This statement should, in addition to the above points, be interpreted as an invitation for employees to suggest to management ways to improve their processes; otherwise, employees cannot do much to improve personal performance beyond following procedures and being alert. Only management provides the elements that control performance, including training, processes, procedures, equipment, work environment, and work assignments. However, Total Quality Management recognizes that the employee who uses the process and faces its problems and weaknesses daily is the one most likely to know how the process or equipment could be changed to improve performance. Managers in the West have been slow to understand this concept, although Deming and his contemporary Dr. Joseph Juran first promoted it in the 1950s. By encouraging these employees in environmentally critical positions to develop and present their ideas, performance improvement with environmental benefits can result.

The intent of Subclause c) is that employees fully understand their roles and responsibilities—their contribution—to conformance with the environmental policy, procedures, and the requirements of the EMS, and their roles and responsibilities in emergency preparedness and response. The organization is held responsible for enabling the relevant employees to gain this understanding through awareness training.

Relevant employees must be aware of the potential consequences of failing to adhere rigorously to established operating procedures. These consequences can include, but are not limited to, environmental impacts that carry significant legal exposure for the organization, damaging publicity that can result in the loss of customers and jobs, cessation of processes and production, and loss of ISO 14000 registration. In some organizations an irresponsible act by an employee can lead to the employee's termination of

> **ISO 14000 INFO**
>
> *Dr. Joseph Juran and Dr. W. Edwards Deming both concluded in the 1950s that approximately 85 percent of an organization's failures were the result of systems controlled by management. Only about 15 percent of the failures could be controlled by workers.[16] Management establishes the systems used by the workers, including the processes, procedures, equipment, training, work environment and culture, and the selection of workers for specific functions. Only management can change any of these system elements. On the other hand, workers can affect results only by the degree to which they adhere to procedures and follow instructions and perform within their capabilities and the limitations placed on them by the organization.*

employment. Subclause d) requires the organization to make employees aware of these consequences through training.

The fourth requirement says the organization must make sure that any employee assigned to a task which has a potential for significant environmental impact is competent to perform that task. ISO does not establish any required level of competence; this is to be determined by the organization. The implication is that for tasks having actual or potential significant environmental impact, the organization must first determine required levels of competence based on formal education, training, and experience. Once this is done, employees already assigned to those tasks should be measured against the requirements. Some will probably be fully competent and qualified for the job; others may require some special training to raise their skills to the required level. Still others may not be sufficiently competent for the job and will have to be replaced by individuals better qualified. The organization determines the meaning of *competency* and ensures that each critical task is done by an employee who meets that standard.

Some organizations use the services of contractors in roles that, for one reason or another, the organization does not want to do itself. It is common for organizations to employ specialized contractors to handle the disposal of hazardous waste. This does not mean that an organization can assign all of its significant impact tasks to contractors and thus evade the responsibility for any impacts, training, and related issues. The organization is still responsible, whether the task is carried out by its employees or by a contractor. Further, ISO 14001, Annex A, Clause A.4.2, states, "The organization should also require that contractors working on its behalf are able to demonstrate that their employees have the requisite training." ISO 14004, Clause 4.3.2.5, says nearly the same thing but stronger and more directly with, " . . . provide evidence that they have the requisite knowledge and skills to perform the work in an 'environmentally responsible manner.'"

Although Clause 4.4.2 does not refer to documenting procedures, the organization is still responsible for documenting its training, awareness, and competence activities and procedures. Clause 4.5.3, linked to this clause, specifically requires that "training records" be maintained. The only way this can be done is through documentation. Training documentation should include at least the following items:

- records of training needs assessments
- task competency requirements
- training procedures
- training plans
- records of training delivered to specific employees

These documented records will be required for the registrar's audits, and they will be invaluable to the organization itself for internal operations and for its defense in case of environmental mishaps.

---

## ISO 14001, Clause 4.4.3: Communication

With regard to its environmental aspects and environmental management system, the organization shall establish and maintain procedures for

a) internal communication between the various levels and functions of the organization;

b) receiving, documenting and responding to relevant communication from external interested parties.

The organization shall consider processes for external communication on its significant environmental aspects and record its decision.

---

### Requirements—Clause 4.4.3

The two requirements of this clause are as follows:

1. The organization must establish and utilize procedures for
   - internal environmental communication, and
   - receiving, documenting, and responding to relevant environmental communications from external interested parties.
2. The organization must consider (explore, look into) various processes for external communication regarding its significant environmental aspects and then record its decision concerning further action.

### Application Information—Clause 4.4.3

The first requirement, stated in Subclause a), to establish and use (maintain) procedures for internal environmental communication is straightforward. The need for open, two-way communication in all aspects of an organization's operations is obvious. Perhaps the need is greater in the environmental area since there are so many legal and regulatory requirements and internally developed requirements that stem from the environmental policy, procedures, and the EMS. Initially these will be new to many employees, so effective communication up and down and laterally should ensure that questions are answered and that understanding is complete and accurate. Even after the EMS is implemented and operating, requirements, technology, processes, procedures, and even employees will change and evolve, continuing the need for the communication channels. The internal environmental communication procedures should address reporting on environmental activities to

- demonstrate management's commitment to the environment and the EMS;
- deal with concerns and questions about the environmental aspects of the organization's activities, products, or services;
- inform relevant employees of all legal and regulatory changes, and changes to the organization's EMS requirements, including the environmental policy, procedures, and documentation;
- raise awareness of the organization's environmental policies;
- ensure that all employees are aware of environmental objectives, targets, programs, and their achievement status;
- publicize results of internal and registrar audits as well as management's review of EMS performance;
- maintain a high level of employee focus on the organization's environmental issues.

The organization should ensure that its internal communication procedures support efficient lateral communication, such as direct communication between employees of different departments. This is frequently a problem, especially in larger, more bureaucratic organizations. Organizations cannot afford the "luxury" of just routing communications up and down through departmental hierarchies. If employees detect an impending environmental impact, they must be able to directly call upon those who can resolve the problem, regardless of departmental affiliation. When a hazardous spill occurs, there had better be a direct communication channel to the response team.

No explicit requirement exists in the clause to document the procedure. In small organizations, documenting an internal communications procedure may not be necessary, but for the vast majority of organizations seeking ISO 14000 registration, documented procedures are the only way to ensure that intended procedures are followed.

The second requirement of the clause, stemming from Subclause b), mandates an organizational procedure to accommodate certain communications with external stakeholders (interested parties). The subclause, first, envisions the organization *receiving* communications related to its environmental aspects from persons or agencies external to the organization and, second, the organization *responding* to those that are relevant. The organization is required to *document* communications received. Several key points to understand are:

1. **Interested parties** can be anyone, any group or any agency. Ordinarily they will be local, state, or federal governmental agencies, recognized environmental groups, or citizens with genuine concerns or questions. Nonetheless, environmental issues

have generated numerous fringe groups and individuals that attempt to judge—and even harass—enterprises *they* think are the causes of acid rain, climatic change, depletion of resources, and other environmental problems. These groups and individuals will need to be included in communications procedures. Organizations would be wise to develop straightforward communication procedures and follow them strictly, regardless of the motivation of the external interested party.

2. The subclause covers only situations in which the external interested party initiates the communication, such as when the organization receives an inquiry.

3. The organization is required to document received communications. A simple approach would be to maintain a file of all written communications received and a log of oral communications, but it is up to the discretion of the organization.

4. The subclause requires the organization to respond to *relevant* communication from external interested parties. It contains no requirement to respond to irrelevant communication or to document it. The formula for determining what is or is not relevant is left up to the organization, but normally communication concerning the organization's environmental aspects and EMS are relevant. From a practical standpoint, it might be unwise to ignore any interested party communication, even if it is irrelevant. For example, suppose that a plant receives an inquiry from a group of local citizens with a genuine concern for their environment. The group believes that the plant might be exposing its neighbors to hazardous chemicals through its highly visible emissions (really steam) to the atmosphere. They write a letter to the plant, expressing their concerns. The organization receives the inquiry and determines that it is irrelevant because no such chemicals are used or produced in its processes. At this point the organization may simply file the letter and take no further action, apparently within the guidelines of ISO 14000. However, the community, in receiving no response, will assume that not only is the plant poisoning its neighbors, it arrogantly refuses to discuss the matter. The situation can only escalate. To avoid misunderstanding, misinterpretation, and possible hostility, it would be far better that the organization view the letter as relevant, since the citizens expressed a concern over an *assumed* environmental aspect and impact. A response could allay the group's concerns, not to mention that one could be easily developed. Moreover, the response would represent an opportunity to capitalize on the organization's environmental efforts by letting the community group know of its environmental policy, EMS, and related activities.

5. It is not clear from Subclause b) whether it is required to document responses to interested parties. The requirement to *respond* to relevant communications places an implicit requirement on the organization to document the response. Auditors will have to see credible evidence of appropriate responses. This can only be done through documentation. It would be a simple solution to maintain a file of inquiries and the corresponding responses. For oral communications, a log of each inquiry and the nature of each response, persons involved, and dates should be sufficient.

The second requirement of Clause 4.4.3 is that the organization *consider* processes to externally communicate information on its significant environmental aspects, and once having considered them, record its decision whether or not to pursue such a process. ISO recognizes that the external communication procedure of Subclause b) is purely reactive. That is, no external communication relative to environmental aspects or the EMS occurs unless it is in response to communication from an external interested party. ISO requires the organization to consider a proactive communication process and to document its decision; i.e., whether to implement the process. An organization declining a proactive procedure should record that fact and supporting reasons.

---

## ISO 14001, Clause 4.4.4:
## Environmental Management System Documentation

The organization shall establish and maintain information, in paper or electronic form, to

    a)  describe the core elements of the management system and their interaction;

    b)  provide direction to related documentation.

---

### Requirements—Clause 4.4.4

This clause has two primary requirements.

1. The organization must establish and maintain (use and keep up-to-date) information, either on paper or in electronic form, that describes the core elements of the EMS and how those elements interact.
2. The organization must establish and maintain (use and keep up-to-date) information, on paper or in electronic form, that provides directions to any related documentation.

### Application Information—Clause 4.4.4

Clause 4.4.4 indirectly mandates an EMS manual, which is analogous to the quality manual for ISO 9000 organizations.

    ISO 14000 requires a documentation system similar to that of ISO 9000. ISO 9000 organizations can easily adapt their documentation systems to accommodate ISO 14000. However, ISO 9000 documentation requirements have always been the biggest stumbling block for registration. Consequently, if the organization is making its initial registration, it can expect to apply a lot of effort in the area of documentation.

    ISO documentation can be viewed in four distinct levels in a hierarchical structure. Refer to Figure 3-6.

*Level 1—Policy Level*    Level 1 refers to the environmental management system manual. The manual describes the environmental philosophy and commitment, environmental issues facing the organization, and plans to avoid environmental problems and improve environmental performance.

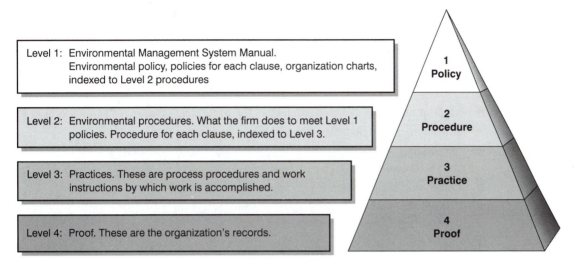

Level 1: Environmental Management System Manual.
Environmental policy, policies for each clause, organization charts,
indexed to Level 2 procedures

Level 2: Environmental procedures. What the firm does to meet Level 1
policies. Procedure for each clause, indexed to Level 3.

Level 3: Practices. These are process procedures and work
instructions by which work is accomplished.

Level 4: Proof. These are the organization's records.

1
Policy

2
Procedure

3
Practice

4
Proof

**Figure 3-6**
Documentation Hierarchy for ISO 14000

The EMS manual will include the environmental policy, with policies responding to each clause of ISO 14001, Section 4, Environmental Management System Requirements. As with ISO 9000, a good starting point for the environmental policy is a clause-by-clause duplication of ISO 14001, Section 4, but with each *shall* changed to *will* and references to *the organization* changed to *we* or *our*, and so forth. For example, where a Section 4 clause states "*The organization shall* establish and maintain. . . . ", the corresponding environmental policy statement would state "*We will* establish and maintain. . . . " instead.

Documentation in the EMS manual will include organization charts and other forms of documents that clearly define core elements of the EMS, how they relate and interact with each other and with the organization, and management responsibility and authority for operating the EMS and each of its elements (Clause 4.4.1).

Although not explicitly required, the organization may want to include in the EMS manual a list of the significant environmental aspects of the organization's activities, products, or services. It is necessary to identify them and keep them up-to-date (Clause 4.3.1), and the organization's environmental objectives and targets will be created to accommodate them, so the EMS manual is the logical place for them.

The EMS manual should list the current environmental objectives and targets. This can be in list form, referring to the actual documentation of the objectives and targets, or it can be the active set of objective and target documentation. In either case they should be updated regularly in order to account for objectives and targets being achieved and removed from the active list (Clause 4.3.3).

It is also a good idea to include applicable legal, regulatory, and other requirements in the EMS manual at least in list form with referral to actual location (Clause 4.3.2).

The EMS manual should include a *list* of environmental procedures corresponding to requirements and to the environmental aspects. The procedures themselves are not located in the manual, because their inclusion would make the manual bulky. The list should provide readers with reference (directions) to actual location of the procedures for anyone needing access.

An EMS manual constructed according to the previous sections will contain these elements:

- Environmental policy reflecting each of the Section 4 clauses.
- Definition of core elements of the EMS and their interrelationships.
- Clear definition of management responsibility and authority for operating the EMS and its elements.
- Current list of significant environmental aspects.
- Current list of environmental objectives and targets (or the actual objective and target documentation).
- Copies of legal, regulatory, and other environmental requirements, or a list of the requirements with references to actual location.
- A list of the organization's procedures related to each of the Section 4 clauses, referencing the ISO 14000 requirement and actual location.

*Level 2—Procedures Level*   Level 2 describes how the organization operates the EMS. As a minimum, it should include procedures to address each requirement of ISO 14001, Section 4. Clearly, only documented procedures can be included in Level 2 of the EMS documentation. One must be careful here, because while the clauses explicitly require some procedures to be documented, others carry only an implicit requirement that can easily be missed until the registrar denotes the error. ISO Clause 4.4.6 a) states that the organization should use documented procedures " . . . to cover situations where their absence could lead to deviations from the environmental policy. . . . " In any organization with more than a handful of people, undocumented procedures can lead to various deviations. For example, the word *documented* does not appear in Clause 4.4.7, which deals with emergency preparedness and response. Unless they are following documented, step-by-step procedures, different people will react to emergencies differently. It is necessary to have a well-planned, documented procedure on which employees can be drilled until they respond consistently; employees should not have to ad-lib in the face of a crisis. Similarly, the procedures should be able to be followed by someone new to the job. Documentation is the only way to achieve these things, and these procedures are also the only ones that can be controlled, an important factor for the organization and for ISO.

Procedure requirements will be found in the following clauses:

| | |
|---|---|
| 4.3.1 | Procedure to identify environmental aspects |
| 4.3.2 | Procedure to identify and have access to legal and other requirements |
| 4.4.2 | Procedures to make employees aware of [various environmental issues] |
| 4.4.3 | Procedures for internal communications |
| 4.4.3 | Procedures for receiving, documenting, and responding to external interested parties |

| 4.4.5 | Procedures for controlling documents |
|-------|--------------------------------------|
| 4.4.5 | Procedures for creation and modification of documents |
| 4.4.6 a) | Procedures for activities where absence could lead to deviations* |
| 4.4.6 b) | Procedures stipulating operating criteria |
| 4.4.6 c) | Procedures related to environmental aspects of goods and services used by the organization |
| 4.4.6 c) | Procedures for communicating relevant procedures and requirements to suppliers and contractors |
| 4.4.7 | Procedures for emergency preparedness and response |
| 4.5.1 | Procedures for monitoring and measuring key operational characteristics* |
| 4.5.1 | Procedure for periodically evaluating compliance* |
| 4.5.2 | Procedure to define responsibility and authority for handling and investigating nonconformance |
| 4.5.3 | Procedures for identification, maintenance, and disposition of environmental records |
| 4.5.4 | Procedure for periodic EMS audits |

While only three of these required procedures are explicitly required to be documented, it is in the best interest of the organization to document all 17. Otherwise, the organization cannot ensure that the procedures are understood and followed by employees. When employees do not follow established procedures, there will always be variability, which can lead to both environmental problems and audit difficulties.

Beyond these required procedures, the organization will probably use several other environmentally related procedures, which should also be included in the Level 2 documentation.

Overall, the procedures describe step-by-step what the organization does to meet the Level 1 policies. Ideally, there should be procedures related to each of the ISO 14001 Section 4 clauses. The Level 1 list of procedures becomes the table of contents for Level 2. In turn, Level 2 should contain a list of the *process procedures* (practices) contained in Level 3.

*Level 3—Practices Level*   Level 3 contains the process procedures (i.e., actual work instructions for activities relevant to the EMS). They represent *what the employees do* in their operational activities. These process procedures, or work instructions, will provide detailed, step-by-step instructions dealing with significant environmental aspects of the organization's activities or for activities required by the Standard. For example, process procedures show how EMS management reviews are to be conducted or how monitoring and measurement equipment is to be calibrated and maintained. Organizations involved in Total Quality Management or ISO 9000 probably will already have these process procedures documented. Organizations that have not documented their process procedures should do so for consistency of operations. The Level 2 list of process procedures (prac-

---

*Explicit requirement to be documented. All others may be implicit requirements.

tices) becomes the table of contents for Level 3. In turn, Level 3 should list the Level 4 forms and records associated with the practices, and where they may be found.

*Level 4—Proof Level*    Level 4 of the EMS documentation is the repository for all forms, records, and the like that represent the objective evidence (proof) that the EMS is, or is not, functioning as it should. In accordance with ISO 14001, these would include, as a minimum, EMS records such as the following:

- Record of decision concerning external communication (Clause 4.4.3)
- Documentation of communication from external interested parties (Clause 4.4.3 b)
- Records related to monitoring and measurement (Clause 4.5.1)
- Records related to maintenance and calibration of monitoring equipment (Clause 4.5.1)
- Records of changes to documented procedures resulting from corrective or preventive action (Clause 4.5.2)
- Training records (Clause 4.5.3)
- Records of results of audits and reviews (Clause 4.5.3)

These records are not only useful to the organization, but they will be the focus of registrar audits, because they are objective evidence that the organization is or is not in **conformance with the ISO 14000 Standard.**

Both the registrar and the organization itself have vested interests in compliance with legal, regulatory, and other environmental requirements. Therefore, it is obvious that in addition to the conformance records listed above, records associated with legal, regulatory, and other requirements should also be maintained. There appears to be no way to avoid this, even though the standard does not explicitly require the maintenance of such records. Under Clause 4.2 the organization commits to prevent pollution and to comply with relevant legislation, regulations, and other requirements. The organization must prove to the registrar's auditors that it is doing so. Since the auditors spend only a few days per year with the organization and do not observe its operations except during those few days, proof can be achieved only through objective evidence, and records will have to be a major part of that evidence. Treat this as a requirement.

An EMS documentation model is shown in Figure 3-7. Such a documentation system, properly responding to the requirements of the standard, will meet the requirements of most organizations.

ISO 14000 is not particular about the media, electronic or paper form, used for documentation systems (see ISO 14001, Clause 4.4.4). An all-electronic documentation system has its advantages. For example, it will be easier to keep current, thereby ensuring that any document used is the latest version. In addition, distribution of new releases is easier, and private documentation files are eliminated. A disadvantage of an electronic documentation system may be less availability because documentation is stored on the computer. Procedures and work instructions must be **readily available** to employees who need them. If the employee who needs a procedure does not have a terminal at his or her work location, the procedure is not readily available. From a practical standpoint, most firms with electronic

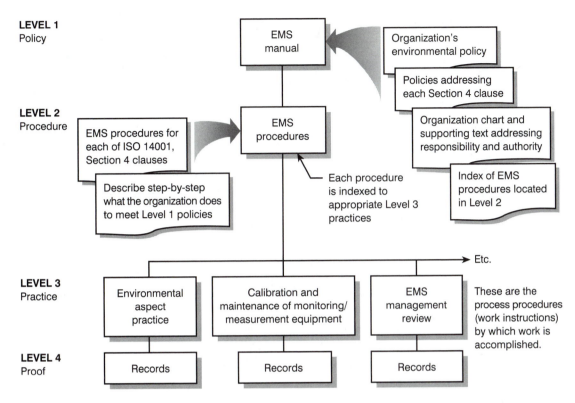

**Figure 3-7**
ISO 14000 EMS Documentation Model

documentation systems find it necessary to print many of the documents for day-to-day use. Even so, the practical utility of an electronic system is worth considering.

Organizations should be aware that ISO 14000 is intended to create an efficient EMS that helps achieve superior and consistent environmental performance; it does not intend to create a bureaucratic or a documentation "monster." Organizations might be wise to take a minimalist approach to ISO 14000 rather than overdoing it on every requirement. Use existing procedures and practices where practical. For example, ISO 9000 organizations will find that many of their procedures fit ISO 14000. New procedures should be as simple as possible, consistent with the job to be done. Keep in mind that it is up to the organization to tell the registrar what it intends to do to conform to the ISO 14000 requirements. Once an organization defines its procedures and the registrar understands those intentions, the registrar will require the organization to conform precisely with those intentions. Simple yet satisfactory procedures will make it easier to adhere to ISO 14000 than complex, difficult procedures. Of course, procedures may be revised as organizations find the need, so there is no compelling reason to start with more detail and complexity than is absolutely needed. ISO 14001, Annex A, Clause A.4.4. Environmental Management System Documentation, states:

The level of detail of the documentation should be sufficient to describe the core elements of the environmental management system and their interaction and provide direction on where to obtain more detailed information on the operation of specific parts of the environmental management system. This documentation may be integrated with documentation of other systems implemented by the organization. It does not have to be in the form of a single manual.

---

# ISO 14001, Clause 4.4.5: Document Control

The organization shall establish and maintain procedures for controlling all documents required by this International Standard to ensure that

a) they can be located;

b) they are periodically reviewed, revised as necessary and approved for adequacy by authorized personnel;

c) the current versions of relevant documents are available at all locations where operations essential to the effective functioning of the environmental management system are performed;

d) obsolete documents are promptly removed from all points of issue and points of use, or otherwise assured against unintended use;

e) any obsolete documents retained for legal and/or knowledge preservation purposes are suitably identified.

Documentation shall be legible, dated (with dates of revision) and readily identifiable, maintained in an orderly manner and retained for a specified period. Procedures and responsibilities shall be established and maintained concerning the creation and modification of the various types of document.

---

### Requirements—Clause 4.4.5

The three requirements of this clause are as follows, working from the end of the clause to the beginning:

1. The organization must establish procedures and responsibility assignments for originating and modifying the various types of documents required.

2. Documents that are created must be legible, dated, readily identifiable, maintained in a structured, orderly manner, and retained for a specified period.

3. The organization must develop and use (maintain) procedures for controlling required ISO 14000 documentation to ensure that

   • documents can be located;

   • documents are periodically reviewed, revised when necessary, and approved for adequacy by authorized persons;

- employees at all locations involved in operations essential to the EMS's effectiveness have appropriate access to relevant documentation;

- obsolete documentation is promptly removed from operational areas and points of issue, or in some manner prevented from unintended use;

- when obsolete documents are retained for legal or historical purposes, they are suitably identified to prevent unintended use.

### Application Information—Clause 4.4.5

Many organizations have a tendency to complicate document control systems. The authors have noted that this is common in organizations that contract with the Department of Defense or are registered to ISO 9000 and/or 14000. This does not best serve the organization, nor is it what ISO 14000 wants. The best document control system is the simplest one that meets the needs of the organization and satisfies the requirements of ISO 14000. The best systems allow for easy, rapid, but controlled, access; ease and timeliness of valid modification and updating; and ready availability of (only) current versions at point of use. The worst systems are slow and ponderous in responding to access or modification and updating needs, and are so complex and rigid that having documentation at points of use where needed is difficult. This leads to all sorts of "workarounds" and subversion of the document control system by employees who are not well served by the system. Workarounds typically subvert the very things the control system is supposed to provide, such as having legitimate control or ensuring that only current versions are in use. ISO does not ask for a complex documentation control system; in fact, they warn against it.[17]

If the organization already has a document control system in place, it should be tested against Clause 4.4.5 requirements. If it satisfies the requirements, continue to use it. If it does not, then modify it to conform. If an organization has no document control system, it must develop one. A good approach is to benchmark a document control system that is successfully used by another organization. In any event, Clause 4.4.5 must be satisfied.

First, the organization must have a document control procedure for originating and modifying the various types of documents required by the standard, including those discussed in the previous section relative to Clause 4.4.4. This procedure should include individuals who are designated responsibility (usually by position title, not name, to avoid revisions when employees move about) for originating or modifying specific categories of documentation. For example, the ISO 14000 Management Representative might be designated as the person responsible for procedures concerned with **record retention.** He or she would be the person to develop, or have developed, procedures spelling out the retention periods of various environmental records. Usually a second person of senior status would also be designated to approve certain categories of procedures. For example, the final approval authority for procedures dealing with records and hazardous waste disposal could be the *vice president of operations*. The approval authority for procedures and work instructions dealing with manufacturing processes that have environmental aspects could be the vice president of manufacturing.

The procedure should provide format instructions for the different kinds of documents. For example, Figure 3-8 illustrates a conforming format for an organization's procedures. There is no requirement to format environmental procedures differently than other kinds of procedures. If the organization already has standard formats for procedures, then it should make sure they conform to the ISO 14000 requirements for **identification**, revisions, dates, requirement, association, and authorized signatures and continue to use them.

Clause 4.4.5 also requires that documentation must (*shall*) be

- legible
- dated
- readily identifiable
- maintained in a structured, orderly manner
- retained for a specified period

The requirement for **legibility** almost seems condescending, but often record files are undecipherable and obscure. Any record that cannot be read or understood by employees or auditors is useless and will not be considered valid, objective evidence. Be sure that handwriting and signatures are readable.

**Dating** is required on records and must be accurate. Dates on records are crucial to verify what happened and on what date. It could literally mean the difference between keeping registration or losing it, or even being charged with an illegal act or being absolved of it. On procedures and work instructions, dates and revision letters or numbers are the only things that identify documents as the correct, current versions. The procedure format of Figure 3-8 requires the date of original issue and the date of revision.

Documents must be readily **identifiable**. Procedures and work instructions should include titles and purpose and should be related to a requiring clause, process, aspect, procedure, work instruction, legislation, or regulation, as appropriate. Note how Figure 3-8 handles this. If the document is a record, make sure that before it is made part of the file it is easily identifiable, even if it means attaching a title page. This is usually not necessary since most documents in the system will be in forms that can be readily identified.

Documents must be maintained (continuously kept) in a **structured, orderly manner**. This means that they cannot simply collect in in-baskets or desk drawers. The document control procedure that the organization is required to establish and use must designate how and where the system stores its documentation. It could all be on a computer, or it could all be in files. More than likely it will be a combination. There is no requirement to have it all at a central location, but it is required that the documentation control procedure specify the location of various documentation.

Good housekeeping principles apply. Files cannot be allowed to deteriorate—there should be no missing documents, irrelevant documents, scrambled order, obsolete documents, or documents kept after their retention period.

**Retention** periods of documents are established by the organization. In some cases there may be periods prescribed by law or regulations, depending on the nature of the organization's operations. Retention requirements must be complied with—even if a document must be kept permanently. Other periods may be at the discretion of the organization, with no outside input. Here the period may be three years, five years, or what-

| XYZ Corporation Procedure | EP 2014 |
|---|---|
| | Procedure No. |
| Title: *Retention of Environmental Records* | **C** |
| Related to: *ISO 14001, Clause 4.5.3* | Revision |

Purpose: *To establish the periods of retention for various categories of records associated with the Environmental Management System.*

Procedure:

(Body of procedure. Use additional pages as required.)

Approved by _____     Date of original issue _____
         *Management Representative*

Approved by _____     Date of revision _____
         *Vice President, Operations*

Page 1 of 1

**Figure 3-8**
Typical Procedure Format

ever the organization determines appropriate. The organization must have a procedure that addresses retention for all the kinds of environmentally related documents in its EMS. In the case of drawings, procedures, and work instructions, which change from time to time, the standard requires that only the latest revision be retained. Certainly only current versions should be available to employees; the document control procedure must address methods to avoid issuing obsolete versions, especially if the organization elects to retain superseded versions.

Clause 4.4.5 requires the organization to establish and use procedures for controlling ISO 14000-related documentation to ensure that

1. documents can be located;
2. documents are periodically reviewed, revised when necessary, and approved by authorized personnel;
3. current versions of relevant documents are available to locations essential to effective functioning of the EMS;
4. obsolete documents are removed from points of use or otherwise prevented from unintended use.
5. retained obsolete documents are suitably marked as such.

As discussed previously, both the procedure and the EMS manual must designate where the various documents are located.

In accordance with item 2, operational documents (procedures, work instructions, drawings, organization charts with lines of responsibility and authority, and other documents that may be affected by the passing of time, changing conditions or operations, and new technology) must be reviewed periodically to ensure that they are up-to-date. An invalid document must be revised and approved. No time period is suggested for such review and revision. Most relevant documents will be revised when changes are made in personnel, operational activities, or lines of responsibility, but the reviews are to catch those documents that do not fall into this category or were overlooked. An annual review should be sufficient. Whatever time period is chosen must be written into the document control procedure.

The procedure must define how employees whose activities are essential to EMS effectiveness are to obtain access to required documentation. In most cases, documentation should be issued in paper form or via computer directly to the relevant operational locations, and it should be continuously available.

The procedure must also address how the documentation system and any outlying stores of documentation are promptly purged of obsolete versions. When a new version of a controlling document, such as a procedure or work instruction, is to be substituted, there must be zero likelihood that an obsolete version may subsequently be used. If the organization elects to retain the obsolete version for its own reasons, the procedure must address the mechanisms that will prevent use of the wrong procedure. This is relatively simple for documentation kept exclusively in electronic form, but for documents kept in paper form, the possibility exists that someone's obsolete private copy will be used instead of the new version.

Finally, Clause 4.4.5 requires that obsolete documents retained for any purpose be clearly identified. Obsolete paper copies should be collected and clearly marked as obsolete so that if they somehow reappear in an operational setting, employees will ask a question before using them. If the obsolete documents are in electronic form, no collection is necessary but they must also be clearly marked as obsolete.

---

## ISO 14001, Clause 4.4.6: Operational Control

The organization shall identify those operations and activities that are associated with the identified significant environmental aspects in line with its policy, objectives and targets. The organization shall plan these activities, including maintenance, in order to ensure that they are carried out under specified conditions by

    a) establishing and maintaining documented procedures to cover situations where their absence could lead to deviations from the environmental policy and the objectives and targets;

    b) stipulating operating criteria in the procedures;

    c) establishing and maintaining procedures related to the identifiable significant environmental aspects of goods and services used by the organization and communicating relevant procedures and requirements to suppliers and contractors.

---

Requirements—Clause 4.4.6

Clause 4.4.6 has a strong linkage to Clauses 4.3.1 and 4.3.3. The two basic requirements are:

1. The organization must (*shall*) identify its *operations* and *activities* that are associated with its significant environmental aspects, its environmental policy, and its targets and objectives.

2. The organization must (*shall*) plan these activities to make sure they are always implemented under specified conditions. The organization must include plans for equipment maintenance.

Subrequirements of the second requirement are:

■ *Documented* procedures must be established and used in any situation where the absence of them could lead to deviations from the policy, targets, or objectives.

■ Procedures must stipulate operating criteria.

■ Procedures must be established and maintained for any purchased or contracted goods or services used by the organization that have identifiable, significant environmental aspects.

■ Relevant procedures must be communicated to the organization's suppliers and contractors.

## Application Information—Clause 4.4.6

At first glance this clause appears to be ISO 14001's catchall for items that do not fit elsewhere. Whether this is the case and whether it was planned this way is unclear. Either way, the organization will have to pay close attention to these requirements. Joseph Cascio cautions readers that this section is the one most apt to get organizations into trouble with auditors if it is misused or overused.[18]

With this clause, ISO 14004 stresses that implementation is accomplished through the establishment and use of operational procedures and controls that ensure achievement of the organization's environmental policy, objectives, and targets.[19]

As we have already seen, Clause 4.3.1 requires that the organization establish and use a procedure for identifying the environmental aspects of its activities, products, and services. Clause 4.3.3 requires that environmental objectives and targets be established, some of which will be tied to the organization's environmental aspects. The first requirement of Clause 4.4.6 specifies that the organization identify its operations and activities that are associated with significant environmental aspects, targets and objectives, and the environmental policy. This will *not* be a list of *all* operations (such as engineering, production, and housekeeping) and *all* activities (such as design, milling, plating, and cleaning). Since the objective is to control those operations and activities that can otherwise cause an environmental impact, result in nonconformance with the environmental policy, or prevent the achievement of stated environmental objectives and targets, only the operations and activities meeting this criteria are to be included.

The second requirement is that the organization *plan* these activities. To *plan* means to devise procedures and/or work instructions that compose a scheme for executing the activities. For each of these activities, a procedure must be developed and employed to ensure that the activities are always carried out as the organization intended (**specified conditions**); this will prevent deviations that could result in environmental impact, nonconformance with the policy, and/or failure to meet objectives and targets.

ISO stipulates that **maintenance** be included in the procedures. This appears to be in the catchall group, but since this requirement appears nowhere else, it is essential that the machines and tools associated with the critical operations and aspects under consideration here have a well-planned, scheduled, comprehensive maintenance program. Obviously, machines that break down or are ill-adjusted can contribute to deviations just as significantly as an operator who does not follow work instructions. The procedure should include any maintenance to be accomplished by the operators, such as lubrication, cleaning, or changing fittings or fixtures; detailed maintenance instructions should probably be included by reference to appropriate manuals.

In order to ensure that the procedures are completed as intended (specified conditions), the organization must do four things:

1. If the organization concludes that using an undocumented (i.e., unwritten) procedure cannot guarantee against deviations from the environmental policy or the

objectives and targets, the procedure must be documented. It is curious that all relevant procedures do not require documentation because it would be wise to do so.

2. Operating criteria must be stipulated in the procedures. The operating criteria are the details and instructions that would normally be included in any process procedure or work instruction. They include things such as equipment to be used, materials required, temperature, speed, proper adapters, nozzles, cutters, and other tools. The criteria also include relevant information on adjustments and step-by-step instructions for the operators. The procedure should provide sufficient information so that there will be no deviation from the standard practice—every operator should do it the same way every time if the instructions are followed.

3. Subclause c) is the Standard's only reference to *suppliers* and *contractors*. It does NOT intend to be a vehicle for "flowing down" the organization's EMS to its suppliers, nor is it intended to be a way to force ISO 14000 on them. It is simply that goods and services purchased from outside suppliers and contractors, if not controlled, can cause deviations in the organization's operations and result in deviations from the policy, its objectives and targets, or even in an environmental impact. For example, suppose the organization uses coal to heat boilers. The organization's practice (process procedure) calls for low-sulfur coal in order to minimize sulfur dioxide emissions. If the coal supplier delivers a high-sulfur variety, and the organization uses it, there could be a deviation from the policy, objectives, and targets and an environmental impact. The organization must establish and use procedures related to the identifiable significant environmental aspects of goods and services provided by suppliers and contractors. These procedures are to be designed to prevent deviations and impacts caused by suppliers and contractors through nonconforming goods or services. Goods and services provided by suppliers or contractors are thus included in the requirement that activities be carried out under specified conditions.

4. Having determined that suppliers or contractors can have an impact on its EMS results, the organization is to communicate relevant procedures and requirements to them.

The operational controls specified by Clause 4.4.6 are actually the Level 3 process procedures and work instructions that the organization determines it requires in order to conform with its environmental policy and to satisfy its objectives and targets. In other words, the operational controls are the procedures and work instructions needed for **implementing the EMS.**

### Requirements—Clause 4.4.7

This clause requires the organization to do three things:

1. Establish and use emergency preparedness and response procedures to
   - identify potential accidents and emergencies;
   - respond to accidents and emergencies;
   - prevent and mitigate associated environmental impacts.

---

## ISO 14001, Clause 4.4.7:
## Emergency Preparedness and Response

The organization shall establish and maintain procedures to identify potential for and respond to accidents and emergency situations, and for preventing and mitigating the environmental impacts that may be associated with them.

The organization shall review and revise, where necessary, its emergency preparedness and response procedures, in particular, after the occurrence of accidents or emergency situations.

The organization shall also periodically test such procedures where practicable.

---

2. Review its emergency preparedness and response procedures and revise them where necessary, especially after accidents or emergency situations.

3. Periodically test these procedures where practicable.

### Application Information—Clause 4.4.7

Organizations must establish emergency preparedness and response plans and procedures that have been tested for adequacy and effectiveness. These plans must cover whatever kinds of environmental emergency situations are conceivable as a result of the organization's operations and activities. Situations might include unintended venting of hazardous gases or liquids, spills, leaks, and fire.

In the United States many firms are required by law to have such plans and procedures. For these organizations, their existing procedures will usually satisfy the ISO 14001 requirements, since the regulatory requirements are more stringent. However, existing procedures should be tested to be sure they satisfy the requirements of this clause. Include the procedures in the EMS manual by reference.

For organizations not already required by law or regulation to have such procedures, the first thing is to establish and use a procedure to identify potential accidents and emergency situations. This should be linked to activities under Clause 4.3.1, which requires the identification of significant environmental aspects. The identification of these significant aspects should have provided the information required to identify potential accidents and emergency situations.

The organization next must develop procedures for responding to the potential accidents and emergency situations. These emergency plans should include:[20]

- emergency organization and responsibilities
- a list of key personnel
- details of emergency services (e.g., fire department, spill cleanup services)
- internal and external communication plans
- actions to be taken in the event of different types of emergencies

- information on hazardous materials, including each material's potential impact on the environment, and measures to be taken in the event of accidental release
- training plans and testing for effectiveness
- equipment and protective clothing required (added to ISO's list)

The organization must also have procedures for preventing and mitigating environmental impacts resulting from accidents and emergency situations. The operational control plans and procedures developed under Clause 4.4.6 are aimed at controlling operational activities to prevent environmental accidents and emergencies in the first place. However, the possibility of accidents and emergency situations always exists. The procedures of this section must address plans to prevent or mitigate environmental impacts resulting from those accidents and emergencies. An example of such a procedure might involve the aspect of leaking chemical drums. The procedure could require that the area in which liquid chemicals are stored have continuous dikes around its perimeter to contain any spilled or leaked chemicals, rather than letting it drain from the storage area into the soil.

The second primary requirement of the clause is that the emergency preparedness and response plans and procedures be reviewed and revised. ISO lets the organization suggest a time period for these reviews, but a one-year cycle is a good starting point. ISO requires that relevant procedures be reviewed after any environmental accident or emergency situation. While the situation is still fresh in everyone's minds, the organization should ask:

- Did we follow the procedure?
- If not, why and how did we deviate?
- What did we do that was right?
- Did we do anything that was wrong?
- If we were to encounter the same situation again, what procedure changes would result in a more effective response?

Actually, the organization does not have to wait for an emergency to ask these questions. They are also appropriate after emergency **drills and simulations**. In any event, the organization may find the opportunity to improve procedures in these postaccident and postdrill activities, and the procedures should be revised accordingly.

Finally, the organization is required, where practicable (feasible), to test its emergency preparedness and response procedures. As indicated, this involves drills and simulations. Organizations should not expect the phrase "where practicable" to offer relief from this requirement. Obviously, actual environmental impacts should not be created just to test the procedures; testing should not result in actual harm. In drills and simulations, water or other materials can be substituted for chemicals and conditions may be simulated.The organization may determine the schedule for testing.

## Requirements Section 5: Measurement and Evaluation

This is EMS Principle 4. ISO 14004 states that "An organization should measure, monitor and evaluate its environmental performance."[21] ISO 14001, Section 4.5, uses the title

**Checking and Corrective Action** *rather than* Measurement and Evaluation. *ISO 14001's* Checking appears to include all of ISO 14004's **measuring, monitoring and evaluating,** but ISO 14001, the Standard, adds **corrective action.** Figure 1, page 4, of ISO 14004 uses both titles in the same box. Although TC 207 did not come to terms with a single title, both work toward the same end. Earlier in this chapter we discussed the Shewhart/Deming roots of ISO's EMS model (*see* Figures 3-1 and 3-2). This section, regardless of its name, represents the third and fourth steps of the Shewhart PDCA Cycle. It intends that the EMS performance be *checked* through *measurement, monitoring, and evaluation* and that action be applied as the need dictates. Although the phrase *corrective action* is used, the intent is clearly more than just corrective action. Actions driven by checking (monitoring, measurement, and evaluation) should include:

- Corrective action—fixing things that were, but are not now, performing as planned.
- Continuous improvement—raising the performance standard of a process that is now doing what was intended, but can be made better still.
- Mitigating impacts—when checking reveals a deviation, action must be taken to avoid or minimize potential or real impacts.

All three of these actions are covered in Clause 4.5.2, which will be explained shortly. The important point is that with a *pure PDCA Cycle approach*, in step 1 the organization would *plan* to do something, and then in step 2 it would implement the plan (*do* it). The material covered in the previous sections has taken us through step 2. In this section the organization will *check* (step 3) the results of the implementation through monitoring, measurement, and evaluation, and it will take appropriate *action* (step 4) in response to the checking. This takes the process back to the beginning, from where the cycle repeats continually.

---

## ISO 14001, Clause 4.5: Checking and Corrective Action

### ISO 14001, Clause 4.5.1: Monitoring and Measurement

The organization shall establish and maintain documented procedures to monitor and measure, on a regular basis, the key characteristics of its operations and activities that can have a significant impact on the environment. This shall include the recording of information to track performance, relevant operational controls and conformance with the organization's environmental objectives and targets.

Monitoring equipment shall be calibrated and maintained and records of this process shall be retained according to the organization's procedures.

The organization shall establish and maintain a documented procedure for periodically evaluating compliance with relevant environmental legislation and regulations.

## Requirements—Clause 4.5.1

Four basic requirements stem from this clause:

1. The organization must establish and use documented procedures for **monitoring and measuring** its operations and activities that can have significant impact on the environment. It must do so in order to track EMS performance, verify operational controls, and assess performance against environmental objectives and targets.

2. The organization must record such information.

3. Monitoring equipment must be calibrated and properly maintained, and records of these activities must be retained.

4. A documented procedure is required for evaluating compliance with applicable laws and regulations.

## Application Information—Clause 4.5.1

With regard to the first requirement, the organization must develop and implement one or more documented procedures to monitor and measure the key characteristics of its operations and activities. Not all operations and activities must be monitored. Rather, the organization must identify its operations and activities that can have a significant impact on the environment (this identification was done in response to Clause 4.3.1). The operations and activities may include:

- Emissions to the atmosphere
- Discharges to soil or water
- Use of water and wastewater discharge
- Use of chemicals and compounds
- Efficiencies of treatments
- Recycling activities
- Hazmat waste generation and storage

Key characteristics of the operation or activity must be monitored and measured. The organization's challenge will be to determine the key characteristics and how to measure them. ANSI/ISO is publishing *Environmental Performance Evaluation Standard 14031*, which may be helpful. An example of an activity's key characteristic and its measurement might be:

| | |
|---|---|
| Activity: | Wastewater discharge |
| Characteristic: | Parameters stipulated in permits and regulations; for example, content of heavy metals, or pH. |
| Measurement: | Sampling of wastewater for heavy metal content, or pH. |

The information derived from monitoring and measuring these key characteristics:

- Is used to track the performance of those operations and activities that can have significant environmental impact.
- Is used to track the performance of the operational controls of those operations and activities.
- Is used to track conformance with the environmental objectives and targets.
- Must be recorded. (This is the second requirement.)

The organization should consider the selection of activities to measure and their key performance indicators to be an ongoing process. As operations evolve and improvements are made to processes, some processes may no longer have significant environmental impact, or new ones may emerge that do. Technological developments may enable the monitoring of new characteristics. Over time, relevant operations and activities and their key characteristics will change, so the organization should review those established under this clause from time to time and update them accordingly.

The third requirement is straightforward, and will be familiar to any organization dealing with federal agencies or ISO 9000. Equipment that is used for monitoring and measurement must be maintained in good working order, and it must be regularly calibrated according to a fixed schedule. The calibration period of most instruments can be determined from the manufacturer's instructions. Although not explicitly required by the clause, a procedure is clearly necessary for controlling calibration and maintenance of monitoring and measurement instrumentation. Documented schedules for preventive maintenance and calibration should be part of that procedure. The clause explicitly requires that records of calibration and maintenance be retained " . . . according to the organization's procedures." This is a link to Clause 4.5.3, Records, which is discussed in a later section. This means that records of instrument repair and calibration activity must be kept, and the organization establishes through its procedures how long they are to be retained, in what form, where they are to be stored, and other pertinent information.

In its environmental policy the organization commits to compliance with all applicable legal and regulatory environmental requirements. The fourth requirement of this clause states that the organization establish and use a *documented* procedure for periodically checking its compliance with the relevant legislation and regulations. The organization, as usual, is free to establish the frequency of its compliance checks and what is to be checked to determine compliance.

It should be clear that ISO 14000 is not a performance standard. Performance standards are established by legislation and regulations and by the organization's own EMS and its objectives and targets. The purpose of this clause is to have the organization test itself against those legal, regulatory, and self-imposed "standards" to determine its areas of success and to identify those areas requiring corrective action and improvement.

Requirements—Clause 4.5.2

The three requirements of this clause are:

1. The organization must establish and implement a procedure for
   - identifying who is responsible for **handling and investigating nonconformances**

---

## ISO 14001, Clause 4.5.2: Nonconformance and Corrective and Preventive Action

The organization shall establish and maintain procedures for defining responsibility and authority for handling and investigating nonconformance, taking action to mitigate any impacts caused and for initiating and completing corrective and preventive action.

Any corrective or preventive action taken to eliminate the causes of actual and potential nonconformances shall be appropriate to the magnitude of problems and commensurate with the environmental impact encountered.

The organization shall implement and record any changes in the documented procedures resulting from corrective and preventive action.

---

- taking mitigating action on nonconformances
- initiating and implementing corrective/preventive action

2. Corrective or preventive actions must be appropriate to the importance of the problem and any resulting environmental impact.

3. Where corrective or preventive action renders a procedure invalid, appropriate changes must be implemented in the procedure.

### Application Information—Clause 4.5.2

The first paragraph of the clause speaks of **nonconformance**, not **noncompliance**. The use of this word seems to indicate that it is not concerned with legal and regulatory noncompliance; in fact, some writers have concluded this.

> This element pertains to any failure to fulfill a requirement of ISO 14001. It does not refer to regulatory noncompliance, although it is frequently misinterpreted as such.[22]

> Nonconformance refers to deviations from the EMS and from requirements of ISO 14001 and should not be confused with noncompliance.[23]

Although this does seem to be the intent of TC 207, it is not certain whether the standard, as written, will be interpreted as intended. As the above quotation indicates, conformance refers to the EMS, and an integral part of the EMS is the organization's environmental policy. ISO 14001 explicitly requires that the environmental policy include a commitment to comply with relevant environmental legislation and regulations. It also requires a commitment to the prevention of pollution. Thus, it seems that if there were a failure to comply with a legal requirement, or if pollution should result from a failure to adhere to a regulation, the result would be *both* noncompliance and nonconformance, since there would be violation of both the legal requirement and the EMS's environmental policy. The organization should discuss this requirement with its registrar and deal with it according to the registrar's particular interpretation.

To satisfy the first requirement of Clause 4.5.2, the organization must develop and implement procedures for:

1. Taking action to mitigate the results of any impacts caused by a nonconformance. This means minimize the extent of the environmental harm done. The action may be containment, cleanup, terminating the offending process, or any other appropriate action.

2. Initiating and completing (implementing) corrective and preventive action. The requirement is not simply to fix whatever went wrong (the corrective action), but also to find the root cause of the problem and a way to eliminate it (the preventive action). This is straight out of the Total Quality Management philosophy of continuous improvement and the teachings of Deming, Juran, and other quality pioneers. It is never enough to merely correct the problem, which restores operation as it was before the problem occurred. Unless the root cause is eliminated, the problem will always return. Procedures for investigating and correcting nonconformance should include these elements:[24]

   - Identifying the cause of the nonconformance.
   - Identifying and implementing the necessary corrective action.
   - Implementing or modifying controls necessary to avoid repetition of the nonconformances.
   - Recording any changes in written procedures resulting from the corrective action.

3. Defining who is responsible for doing the above and delineating the authority afforded to them. This can be in the form of an environmentally oriented organization chart, job description, or anything else that clearly and effectively defines the responsibility and authority.

Although there is no explicit requirement for this procedure to be documented, except from a practical point of view, it is difficult to imagine any other way to satisfy the requirements.

The second requirement of the clause, making any corrective or preventive action "appropriate to the magnitude of problems and commensurate with the environmental impact encountered." seems to be stating the obvious. Joseph Cascio says, "An appropriate action corrects the nonconformance and, at the same time, is not an overkill."[25] When problems arise, they must be dealt with, even if they are not at the top of the priority list. Clearly, organizations should not assign the entire workforce to solve a single problem that could be handled by a few well-chosen employees. On the other hand, significant problems should have more attention, especially from management. The key point to remember is that nonconformances must be corrected, quickly and rationally using appropriate resources.

The third requirement of the clause states that procedures or work instructions must be revised to take corrective and preventive actions into account. If this is not done, the procedures and the corrective and preventive actions will be at odds, and there will be more nonconformances.

## ISO 14001, Clause 4.5.3: Records

The organization shall establish and maintain procedures for the identification, maintenance and disposition of environmental records. These records shall include training records and the results of audits and reviews.

Environmental records shall be legible, identifiable and traceable to the activity, product or service involved. Environmental records shall be stored and maintained in such a way that they are readily retrievable and protected against damage, deterioration or loss. Their retention times shall be established and recorded.

Records shall be maintained, as appropriate to the system and to the organization, to demonstrate conformance to the requirements of this International Standard.

### Requirements—Clause 4.5.3

The four basic requirements under this clause are:

1. The organization must store and maintain (protect) records that are necessary to demonstrate conformance to ISO 14000 requirements, including training records and the results of audits and management reviews.
2. Records must be:
   - legible
   - identifiable
   - traceable to the relevant activity, product, or service
   - readily retrievable
   - protected against damage, deterioration, or loss
3. Record retention times must be established.
4. The organization must establish and use procedures to accomplish all of the above.

### Application Information—Clause 4.5.3

Clause 4.5.3 establishes the requirement to retain environmental records even where clauses directly related to a specific activity may not require it. For example, Clause 4.4.2, while addressing training, contains no requirement for training records, although that requirement is explicitly called out in Clause 4.5.3. The same is true for audits and management reviews (Clauses 4.5.4 and 4.6, respectively). Clause 4.5.3 is linked to these three clauses by explicit requirements for retaining specific records. It is also linked to several other clauses, including:

- 4.3.1 Environmental aspects—for the listing of aspects and impacts.
- 4.3.2 Legal and other requirements—for legislative and regulatory requirements, permits, etc.

- 4.4.6 Operational control—for supplier and contractor information.
- 4.5.1 Monitoring and measurement—for inspection, calibration, and maintenance activity.
- 4.5.1 Monitoring and measurement—for data obtained.
- 4.5.2 Nonconformance and corrective and preventive action—for details of nonconformances, incidents, complaints, and follow-up action.

In addition to these, there will likely be several more categories of records to be maintained, including those required by legislative, regulatory, and other requirements. Some examples include:

- information on applicable laws, regulations, and other requirements
- complaint records and responses
- objective and target achievement data
- process information (procedures, work instructions, etc.)
- product information (aspects, impacts, disposal, etc.)
- information on emergency preparedness and response (procedures, response data)
- government agency inspection results and actions
- waste generation
- chemical usage
- Hazmat safety data sheets

Records should include those required by law and ISO 14001 as well as all other records needed for the implementation and ongoing effectiveness of the EMS.

In reference to the first requirement, the organization must store and maintain records that are necessary to demonstrate conformance with ISO 14000 requirements. Remember that since the environmental policy must commit to compliance with legal, regulatory, and other requirements, relevant records must be considered part of the package to demonstrate conformance with ISO 14000 and the EMS. In addition to records of training, audits, and management reviews, the organization should also maintain at least those records shown in the previous list. Keep in mind that these records will be used by the registrar's auditors as well as by employees working within the EMS.

The second requirement makes specific demands related to the records. They are:

- First, they must be legible. This was discussed under Clause 4.4.5 and is just common sense.
- Second, records must be identifiable. This is common sense, too. Records must be identifiable by category, date, and other pertinent features.
- Third, they must be traceable to the activity, product, or service to which they apply. This is an important point. One needs to know if a record applies to one process or another, is related to product A or B, and so forth.

- Fourth, records must be readily retrievable. Both the organization's employees and the registrar's auditors will have to be able to retrieve records quickly. Records may also have to be made available to inspectors from governmental agencies. A convenient storage and retrieval system is needed.

- Fifth, the records storage and retrieval system must protect the records from damage, deterioration, or loss. No organization can afford the loss of its records. Continued registration demands the ready availability of records. In addition, records may be required to avert fines or severe penalties. Records that are damaged or deteriorated may not be adequate.

The third requirement is that retention times be established. The time period a particular record must be retained depends largely upon the nature of the record. Some types of environmental records should be kept indefinitely for legal reasons and to serve as protection in case of liability. Others may have no useful value after a few years. The organization must establish the retention periods for all categories of environmental records, considering legal and regulatory requirements. Organizations may want to be conservative initially by setting retention periods longer than believed necessary. The periods can be adjusted later if shorter periods will suffice.

The fourth requirement is that the organization establish and use procedures to accomplish everything we have talked about under this clause. That is, the procedures must detail what records are to be retained, where they are to be kept, and for how long. They must indicate how the organization assures that the records are legible, identifiable, traceable, retrievable, and protected. If the organization is already registered to ISO 9000, then including the environmental records under the existing records procedures should not be difficult. Most organizations of any size already use systems for storage and retrieval of documents and records. These systems can be readily adapted. For organizations that do not have such procedures, an organization that has an efficient storage and retrieval system for its records might be identified and used for benchmarking.

## Requirements—Clause 4.5.4

This clause requires two things:

1. The organization must establish and operate a *program* of periodic audits of the EMS designed to

   - determine and ensure conformance with ISO 14001 and whether it has been properly implemented and used; and

   - provide information about audit results to management as part of their continuous improvement program.

   The program's schedule of audits, determination of activities to be audited, and the frequencies of audits must be based on the significance of environmental aspects, impacts of the activities, and past audit performance of activities.

2. The organization must develop and implement procedures necessary for carrying out the audit program. The audit procedure must cover the following:

---

## ISO 14001, Clause 4.5.4:
## Environmental Management System Audit

The organization shall establish and maintain (a) program(s) and procedures for periodic environmental management system audits to be carried out, in order to

    a) determine whether or not the environmental management system

        1) conforms to planned arrangements for environmental management including the requirements of this International Standard; and

        2) has been properly implemented and maintained; and

    b) provide information on the results of audits to management.

The organization's audit program, including any schedule shall be based on the environmental importance of the activity concerned and the results of previous audits. In order to be comprehensive, the audit procedures shall cover the audit scope, frequency and methodologies, as well as the responsibilities and requirements for conducting audits and reporting results.

---

- Audit scope
- Frequency of audit
- Audit methodologies
- Responsibilities and requirements for auditing and reporting results

### Application Information—Clause 4.5.4

The first requirement pertains to the periodic environmental audit program. Clause 4.5.4 is concerned with a program of audits of the EMS and not with compliance to laws and regulations or absolute environmental performance. It is interested in determining if all the requirements of ISO 14001 are being addressed through the EMS and if all the self-imposed requirements of the EMS are being met through implementation of the EMS. In addition, although it does not say so, this clause clearly discusses **internal** or **self-audits**. For example, the registrar would not use the audit program (or procedure) required here for conducting its registration audit or surveillance audits. Under this clause, the audit criteria is set by the organization. This means the auditors will compare what the organization said it would do with what it is actually doing under its implemented EMS. The auditors will probably be employees, although there is nothing to prevent the organization from engaging third-party auditors if it so chooses. Upon concluding the audit, a report is to be given to management. Interestingly, ISO 14001 makes no explicit requirement to document the report, although in reality there is no other practical way to go about it, especially in light of Clause 4.5.3's requirement for records of audit results.

Not all activities of an organization may require environmental audits—only those associated with the EMS. Of these, some offer little or no potential for environmental impact, while others may have a significant potential for environmental impact. The latter are to be scheduled for more frequent audits than those with lesser environmental roles. Other criteria to be considered in setting the frequency of audits for the activities include the activities' track record and recent audit results. Activities that have few environmental problems would be scheduled less often than activities which have many. Activities with no audit nonconformances in recent audits should be audited less often than those that have more nonconformances.

The second requirement mandates procedures for periodic EMS audits. The organization must develop and use procedures for implementing the EMS audit program just explained. The procedure(s) must cover four areas:

1. **The audit scope.** The scope is to be limited to the requirements set by the EMS and should not concern itself with absolute environmental performance.

2. **Frequency of audit.** The organization must determine and specify audit frequency. Probably most organizations use schemes similar to ISO 9000 organizations. Following the original registration audit, the registrar performs surveillance audits at six-month intervals. Most organizations schedule internal audits between surveillance audits, which provides a check on any action items left by the surveillance audit and allows them to discover problems before the next surveillance audit. At least initially, a general six-month interval to audit all relevant activities might be good, staggered between the registrar surveillance audits. Using this technique, the period may be extended for activities that have few problems or for activities with little potential for environmental impact.

3. **Audit methodologies.** The procedures should define whether the audits are to be performed by employees or by outside auditors. If they are to be internally manned, then mechanisms must be built into the procedures to assure the objectivity of auditors. For example, internal auditors should audit only activities of which they are independent; independence promotes objectivity. Methodologies for collecting evidence will be considered in this section. Audit procedures will be guided by ISO 14010, 14011, and 14012.

4. **Responsibilities and requirements for conducting audits and reporting results.** This section considers the assignment of responsibility for conducting audits and the requirements for reporting results to management. It should include an organization chart or a similar device. Report formats should also be included. The organization will use ISO 14010, 14011, and 14012 for guidance.

## Requirements Section 6: Review and Improvement

In previous sections we discussed four of the five environmental management system principles listed in ISO 14004, Clause 4. They were as follows:

■   Commitment and policy—management's commitment to its environmental policy, including prevention of pollution; compliance with relevant laws, regulations, and its own environmental requirements; and continuous improvement of its EMS.

- Planning—in which the organization's environmental aspects and impacts are identified, corresponding objectives and targets are established, and plans are developed for their fulfillment.

- Implementation—in which awareness and competence training is provided; the EMS's structure with roles, responsibilities, and authorities is defined; internal and external communication systems established; environmental documentation is developed, along with systems for control of documents; a system for control of operations related to environmental aspects is established; and an emergency preparedness and response system is implemented.

- Measurement and evaluation—in which a system for monitoring and measuring key environmental characteristics, dealing with nonconformances, taking corrective and preventive actions, keeping environmental records, and establishing a program of EMS audits is established.

To this point, then, the EMS system is presumably doing what management intended it to do. How does management know how effectively it is working, if there are specific areas of substandard performance, or that a particular element might need attention and improvement? Management can know only if it deliberately sets up a series of reviews of the data comprising all the previously discussed sections, covering all of the elements of the EMS. This brings us to the fifth and final EMS principle. Principle 5 is *Review and Improvement*. It states, "An organization should review and continually improve its environmental management system, with the objective of improving its overall environmental performance."

This element of the EMS **demonstrates** management's real commitment to improving environmental performance continually and permanently. It is one thing to write (or have written) a high-minded environmental policy and oversee the work expended through implementation of the EMS; it is quite another to maintain the continuous high level of interest and participation over time that is necessary in order to meet the requirements of this section. The concept cannot be overstated—if management does not carry out its responsibilities under this section of ISO 14000, the whole effort will have been a failure.

---

## ISO 14001, Clause 4.6: Management Review

The organization's top management shall, at intervals that it determines, review the environmental management system, to ensure its continuing suitability, adequacy and effectiveness. The management review process shall ensure that the necessary information is collected to allow management to carry out this evaluation. This review shall be documented.

The management review shall address the possible need for changes to policy, objectives and other elements of the environmental management system, in light of environmental management system audit results, changing circumstances and the commitment to continual improvement.

## Requirements—Clause 4.6

The four basic requirements of this clause are:

1. Top management must periodically review the EMS.
2. Management must define what information it requires for such reviews and ensure that the information is collected.
3. Reviews must be documented.
4. Reviews must address possible needs for changes in any of the EMS elements.

## Application Information—Clause 4.6

The intent of Clause 4.6 is that top management involve itself in the operation of the EMS continuous improvement process. This is done by using feedback data from all the EMS elements as a primary input to management's *checking* function of the EMS model (Figure 3-2), or, if you prefer, the *check* function of the PDCA Cycle (Figure 3-1). Management is to evaluate the feedback data and make corrections or improvements where indicated. This is the vehicle that will enable ISO 14000 organizations to continuously improve environmental performance.

No requirement exists for the organization to develop and maintain a procedure for management review of the EMS. It would be a good idea, however, because there will be so many different activities involved that without a documented procedure of the process, it will be difficult for management to get what it wants. A documented procedure should be developed around the requirements that follow.

The first requirement is for top management to periodically review the organization's EMS to ensure that it

- continues to be suitable in light of changing conditions,
- remains adequate in light of current expectations, and
- is effective in producing the desired results and performance.

Management determines the appropriate review interval. A monthly review should be considered; intervals longer than a month tend to result in the data becoming stale and interest in the issues waning. No requirement exists for the entire EMS and all of its elements to be reviewed at once. It is possible, if the organization wishes, to review certain elements at one time and others at a different time. The point is, There should be a schedule of reviews covering all elements.

The second requirement of the clause states that the "management review process" should ensure that information necessary for reviewing the EMS is collected. Management has to tell the employees what information it wants for the review process. The best way to do this is through a documented procedure, similar to that just discussed. Most of the information for the reviews will come from the tracking of key characteristics of the EMS, progress and issues on objectives and targets, and audits, both internal and external.

The review of the EMS should be broad enough in scope to address the environmental dimensions of all activities, products or services of the organization, including their impact on financial performance and possibly competitive position.[26]

In particular, the EMS reviews should include: [27]

- review of environmental objectives and targets.
- review of environmental performance against legal, regulatory, and other requirements.
- evaluation of effectiveness of the EMS's elements.
- evaluation of continued suitability of the environmental policy in light of
  - changing legislation,
  - changing expectations,
  - changing requirements of interested parties,
  - changes in products, services, or activities of the organization,
  - new technology,
  - lessons learned,
  - market preferences and expectations, and
  - effectiveness of reporting and communication.

The third requirement of Clause 4.6 is simple: Management must keep records of its EMS reviews. These records will be subject to the organization's procedures for its environmental records under Clause 4.5.3. The records will be a subject for review by internal and external auditors.

In keeping with the fundamental intent of the clause, the fourth requirement asks management to address the "possible need for changes to policy, objectives and other elements" of the EMS in its reviews. No EMS, regardless of how effective it has been or how flawlessly it has been implemented, will be adequate forever. When circumstances change or when the organization changes its activities or products, corresponding EMS changes will be required. Even if circumstances do not change or even if the organization continues with its same activities and products, neither of which is likely, changes to the EMS will still be required. The organization will find ways to improve elements of the EMS, and according to its commitment to continual improvement, management will have to be continually on the lookout for EMS improvement opportunities.

---

### ISO 14000 INFO

*The concept of continual improvement is embodied in the EMS. It is achieved by continually evaluating the environmental performance of the EMS against its environmental policies, objectives, and targets for the purpose of identifying opportunities for improvement.*
   *ISO 14004, Clause 4.5.3*

The continual improvement process should[28]

- compare actual results with desired results of objectives and targets, and legal, regulatory, and other requirements,
- identify opportunities to improve environmental performance,
- identify root causes of nonconformances or deficiencies,
- plan and implement action to eliminate root causes,
- verify effectiveness of action, and
- document resulting changes in procedures, processes, etc.

This is the classic continuous improvement model from Total Quality Management (TQM). Organizations already involved in TQM will find this an easy element to implement; those who have not adopted TQM will probably struggle with this concept until employees at all levels see that the results are worth the effort. Change clearly takes effort, and it has a tendency to concern employees and management alike; it is easy to become comfortable with a stable, unchanging environment. Unfortunately, though, an unchanging organization often is adversely affected and inevitably falls behind its competitors who, through change, press onward to greater performance and efficiency and improve their competitive position. In our competitive world, neglecting change and continual improvement leads to economic disaster and eliminates any possibility of better environmental performance, which the planet so badly needs.

## SUMMARY

1. The relationship of ISO 14000 to legislative and regulatory requirements is as follows: Legislative and regulatory agencies make environmental laws and regulations; ISO does not. ISO 14000 requires only that registered organizations comply with all applicable laws and regulations through a structured environmental management system.

2. ISO 14001, Clause 4.1: General Requirements, mandates that an organization establish and maintain an EMS that meets all specified requirements in the following areas: environmental policy (Clause 4.2), planning (Clause 4.3), implementation and operation (Clause 4.4), checking and corrective action (Clause 4.5), and management review (Clause 4.6).

3. ISO 14001, Clause 4.2: Environmental Policy, requires that an organization define its policy and ensure commitment to its EMS. An environmental policy establishes an overall sense of direction for an organization and the principles of action by which it will be judged. The policy must contain at least the following elements: (1) commitment to continual improvement, (2) commitment to the prevention of pollution, and (3) commitment to comply with all applicable environmental legislation, regulations, and other requirements to which the organization subscribes.

4. ISO 14001, Clause 4.3: Planning, requires at least the following planning components: identification of environmental aspects and evaluation of associated environ-

mental impacts, legal requirements, environmental policy, internal performance criteria, environmental objectives and targets, and environmental plans and management program.

5. ISO 14001, Clause 4.4: Implementation and Operation, requires organizations to address the following seven elements of an environmental management system: organizational structure and responsibilities; environmental training, awareness, and competence; communication; EMS documentation; control of documentation; control of operational activities; and emergency preparedness and response.

6. ISO 14001, Clause 4.5: Checking and Corrective Action/Measurement and Evaluation, requires that EMS performance be checked through measurement, monitoring, and evaluation. Actions driven by checking should include corrective action, continuous improvement, and mitigating impacts.

7. ISO 14001, Clause 4.6: Management Review, requires organizations to review and continually improve their EMS, with the objective of improving its overall performance.

## KEY CONCEPTS

Checking and corrective action

Commitment

Communication—internal, external

Competitive advantage

Compliance to legal and regulatory requirements

Conformance to the Standard

Continual improvement

Control of documentation

Control of operational activities

Drills and simulations

Emergency preparedness and response

EMS documentation

Environmental aspects

Environmental management programs

Environmental management system (EMS)

Environmental objectives and targets

Environmental policy

Environmental training, awareness, and competence

Handling and investigating nonconformances

Implementing and operating the EMS

Interested parties

Management Representative

Management review

Monitoring and measuring

Organizational structure and responsibilities

PDCA Cycle

Planning for the EMS

Practices and procedures

Prevention of pollution

Record retention

## REVIEW QUESTIONS

1. Explain the relationship between ISO 14000 and national legislative and regulatory agencies.

2. What is an EMS?

3. Why does ISO stress the need for commitment by top management?

4. What is an environmental policy?

5. List five concerns that should be considered in an environmental policy.

6. Explain the concept of continuous improvement as it relates to ISO 14000.

7. Differentiate between an objective and a target.

8. What does ISO mean by *documenting* the environmental policy?

9. List six elements that should be part of an organization's environmental plan.

10. Name the four key areas an organization should review in order to establish its current environmental position.

11. In addition to legal requirements and significant environmental aspects, what other concerns should a company consider (Clause 4.3.3)?

12. Explain the concept of a *program* as it relates to objectives and targets.

13. Explain the concept of *implementation* as it relates to ISO 14000.

14. What is the *Management Representative* and what responsibilities correspond to the position?

15. How should an organization determine training needs relating to ISO 14000?

16. List the contents of a good training documentation package.

17. Explain the four levels of documentation (hierarchy) for ISO 14000.

18. What are operating criteria?

19. Actions driven by checking (monitoring, measurement, and evaluation) should include certain elements. List three of those elements.

20. How does an organization know how long to retain a given record?

## CRITICAL-THINKING PROBLEMS

The following activities may be assigned as individual, group, or discussion activities to be completed in class or out of class.

1. You recommended that your company pursue ISO 14000 registration. The company's CEO responded, "Why bother? What's it going to do for us?" Respond to this question with a concise, but comprehensive, and convincing rationale.

2. Using your home as the subject of this activity, identify two environmental aspects in your home and potential environmental impacts associated with them. Establish hypothetical environmental objectives and targets for the impacts.

3. You have been given responsibility for promoting environmental awareness among employees in your company. How will you proceed?

4. Use the *Environmental Objective and Target Action Plan* in Figure 3-4 with this activity. Complete the form for the objectives/targets you identified in Question 2 of this section.

5. Your company needs a Management Representative to be its EMS czar. The top managers must decide whether to create a new full-time position or add these duties to an existing position. Which option do you recommend and why?

6. You are your company's EMS czar. You are preparing to identify the workforce's training needs with regard to ISO 14000 when your boss says, "Forget it. Do only the minimum you have to do to satisfy the training requirement." Prepare a justification for a comprehensive and thorough approach to training.

7. You will be responsible for coordinating the development of your company's EMS manual. What elements should be included in the manual? Your boss wants process procedures (work instructions) to be included. What should you tell him?

8. Refer to Figure 3-8. Complete the procedure narrative. Decide how you would go about satisfying the stated purpose, and then document your procedures in writing.

9. Call several companies in your community and ask the safety or environmental manager if the organization has a plan for accidents and emergencies. Schedule an appointment to review the plan of at least one company. Critique the plan. Does it cover all of the necessary elements? If not, what is missing?

10. If a company meets all requirements of ISO 14000 and the applicable state and federal regulations, why should it concern itself with continuous improvement?

========= DISCUSSION CASE =========

## Award-Winning Environmental Program

"Circuits Engineering Inc. (CEI), a printed circuit board manufacturer in Bothell, Washington, has gone to great lengths to promote its partnership with the environment, regulatory agencies, its customers, and its employees.

"Through its in-house Environmental Awareness Team and Waste Prevention Challenge Program, CEI has made big strides. During the period of 1996 to 1998, CEI decreased its annual hazardous waste generation by over 45 percent and decreased its average daily generation of industrial water by more than 11 percent. In addition, CEI currently recycles more than 70 percent of its total solid waste stream.

"CEI intends to continually improve its environmental performance.

"CEI intends to reduce the total chemicals it stores on site by 90 percent. The company plans to accomplish this by changing from a current process that uses copper, formaldehyde and cyanide to an alternate process that uses carbon as a main constituent.

"In light of the manufacturer's accomplishments, Washington State Department of Ecology honored CEI in 1997 as having the best small business recycling program in Washington.

"The company is an active partner in the U.S. Environmental Protection Agency's WasteWise Program and the Energy Star Small Business Program."[29]

### Discussion Questions

With the commitment CEI has made and the results it has achieved, is there any reason why this company should pursue ISO 14000 registration? Why or why not?

## ENDNOTES

1. ANSI/ISO 14004-1996, p. vii.

2. Ibid., p. 6.

3. Ibid., p. 5.

4. Mary Walton, *The Deming Management Method* (New York: The Putnam Publishing Group, 1986), pp. 86-88.

5. ANSI/ISO 14004-1996, p. 8.

6. Ibid., pp. 8-10.

7. ANSI/ISO 14001, A.3.1, p. 7.

8. Ibid., A.3.2, p. 7.

9. ANSI/ISO 14004-1996, 4.2.3, p. 9.

10. Ibid., 4.2.5, p. 11.

11. Ibid.

12. Cascio, Woodside, and Mitchell, *ISO 14000 Guide: The New International Environmental Management Standard* (New York: McGraw-Hill, 1996), p. 110.

13. Ibid., p. 169.

14. Tom Tibor, *ISO 14000: A Guide to the New Environmental Management Standards* (Chicago: Irwin Professional Publishing, 1996), p. 60.

15. ANSI/ISO 14004-1996, 4.3, p. 13.

16. Peter R. Scholtes, *Total Quality Management* (Southfield, MI: Peter Scholtes, Inc., 1991), pp. 1-12.

17. ANSI/ISO 14001-1996, Annex A, Clause A.4.5, p. 9.

18. Joseph Cascio, *The ISO 14000 Handbook* (Milwaukee: ASQ Quality Press, 1996), p. 179.

19. ANSI/ISO 14004-1996, 4.3.3.3, p. 20.

20. Ibid., Clause 4.3.3.4, p. 21.

21. Ibid., Clause 4, p. 3.

22. Marilyn R. Block, *Implementing ISO 14001* (Milwaukee: ASQ Quality Press, 1997), p. 71.

23. Cascio, Woodside, and Mitchell, p. 138.

24. ANSI/ISO 14001-1996, Annex A, Clause A.5.2, p. 9.

25. Joseph Cascio, p. 184.

26. ANSI/ISO 14004-1996, Clause 4.5.2, p. 23.

27. Adapted from ANSI/ISO 14004-1996, Clause 4.5.2, p. 23.

28. Ibid., Clause 4.5.3, p. 24.

29. Angela Neville, "Facilities of the Year," *Environmental Protection*, vol. 9, no. 12, December 1998, p. 22.

# The Environmental Management System (EMS)

## DEFINITION—WHAT IS AN EMS?

In chapter 2 under the section Language of ISO 14000, the ISO 14001 definition of an **EMS** was quoted as "the part of the overall management system that includes organizational structure, planning activities, responsibilities, practices, procedures, processes and resources for developing, implementing, achieving, reviewing and maintaining the environmental policy." The definition, though, seems to fall apart, as is true with many ISO definitions. It seems ISO is saying that the EMS enables the development of the environmental policy, but the environmental policy itself is the focal point of any EMS. By definition, then, at the point in time when the environmental policy is being developed, there can be no EMS, since for it to exist, the environmental policy must already exist. The purpose of the EMS, as stated in ISO 14001, Annex A, Section A.1, is:

> The environmental management system provides a structured process for the achievement of continual improvement. . . . "

Obviously, the intent is that through implementation of an EMS, the organization will be able to continually improve its environmental performance. In that perspective, the EMS is "the part of the overall management system that includes organizational structure, planning activities, responsibilities, practices, procedures, processes, and resources for achieving sound environmental performance and for reviewing, maintaining, and improving performance under the environmental policy." Under this definition, both the development and initial implementation of the environmental policy become prerequisites of the EMS.

While the latter definition seems more accurate, for all practical purposes, both are acceptable. Remember that the EMS is the part of the organization's overall management system that addresses the environmental aspects of its operations, enabling sound environmental performance and leading to **continual improvement**.

## EMS REFERENCES IN ISO 14000

References to an environmental management system relative to constructing, implementing, and operating the EMS may be found in:

- ISO 14001-1996

| | |
|---|---|
| Introduction | Background information |
| Clause 1 | Establishing scope and applicability |
| Clause 3 | Definitions applicable to the EMS |
| Clause 4 | (All of section 4) Establishing EMS requirements |
| Annex A | Providing information relative to each of the Clause 4 subclauses |

- ISO 14004-1996

| | |
|---|---|
| Introduction | Background information |
| Introduction 0.1 | Overview and background information for an EMS |
| Introduction 0.2 | Benefits of having an EMS |
| Clause 1 | Scope of the guidelines contained in ISO 14004 |
| Clause 3 | Definitions relative to the EMS guidelines |
| Clause 4 | EMS principles and elements—offers implementation guidance and practical help on a paragraph-by-paragraph basis for ISO 14001, Clause 4 (all). |

Note: Users would have found it less confusing and easier to follow if the ISO 14001 clause numbering scheme were used in ISO 14004. For example, ISO 14001, Clause 4.2, is concerned with environmental policy, whereas ISO 14004, Clause 4.2, is concerned with planning. Fortunately, though, the clauses of ISO 14004 do follow the same order as those of ISO 14001, so keeping track is not difficult.

# MANAGEMENT RESPONSIBILITY

There are two levels of management responsibilities in connection with ISO 14000. The first level relates to the responsibilities specified in the Section 4 clauses of ISO 14001. The second level relates to the responsibilities that are not specified but are necessary to assure that the EMS functions effectively.

Regarding the specified responsibilities, any Section 4 clause that states something like "the organization shall establish. . . . " can be interpreted as an implicit management responsibility. Even though the statement does not explicitly mention management, only management has the authority and power to establish. Such implicit management responsibilities are as follows:

| Clause | Management's Implicit Responsibilities to Establish and Maintain |
|---|---|
| 4.1 | The EMS |
| 4.3.1 | Procedures regarding environmental aspects and impacts, and currency of |
| 4.3.2 | Procedures regarding legal and other requirements |
| 4.3.3 | Environmental objectives and targets |
| 4.3.4 | Programs for achieving environmental objectives and targets |
| 4.4.2 | Procedures regarding **awareness and competence training** |
| 4.4.3 | Procedures for internal and external environmental communication |
| 4.4.4 | Information and documentation regarding the EMS |
| 4.4.5 | Procedures for controlling EMS documentation |
| 4.4.6 | Documented procedures for controlling critical operations |
| 4.4.6 | Procedures regarding critical goods and services used by the organization |
| 4.4.7 | Procedures regarding emergency preparedness and response |
| 4.5.1 | Documented procedures regarding monitoring and measuring key operational characteristics |
| 4.5.1 | Documented procedures for periodic compliance evaluation |
| 4.5.2 | Procedures for defining responsibility and authority for corrective and preventive action |
| 4.5.3 | Procedures for identification, maintenance, and disposition of records |
| 4.5.4 | Program and procedures for periodic EMS audits |

---

**ISO 14000 INFO**

*"To ensure success, an early step in developing or improving an EMS involves obtaining commitment from the top management of the organization in order to improve environmental management of its activities, products or services. The ongoing commitment and leadership of the top management are crucial."*
   *ANSI/ISO 14004-1996, Clause 4.1.2*

In addition to the "establish and maintain" category, the Section 4 clauses contain several other implicit management responsibilities. They are as follows:

| Clause | Management's Implicit Responsibilities |
|--------|----------------------------------------|
| 4.4.1 | Define, document, and communicate environmental roles and responsibilities |
| 4.4.1 | Provide essential resources to implement and control the EMS |
| 4.4.2 | Identify training needs and provide appropriate training |
| 4.4.6 | Identify operations and activities associated with significant aspects |
| 4.4.6 | Plan relevant activities to assure control of these operations |

ISO 14001, Section 4, also has some explicit management requirements, as follows. These requirements begin with "Top management shall. . . . "

| Clause | Management's Explicit Responsibilities |
|--------|----------------------------------------|
| 4.2 | Define the environmental policy |
| 4.2 a) | Ensure that the environmental policy is appropriate |
| 4.2 b) | Commit to continual improvement and prevention of pollution |
| 4.2 c) | Commit to comply with legal and other requirements |
| 4.2 d) | Ensure that the environmental policy accommodates setting and reviewing of environmental objectives and targets |
| 4.2 e) | Ensure that the environmental policy is documented, implemented, used, and communicated to all employees |
| 4.2 f) | Ensure that the environmental policy is available to the public |
| 4.4.1 | Appoint an EMS Management Representative; assign the representative's roles, responsibilities, and authority |
| 4.6 | Conduct periodic reviews of the EMS to ensure continuing suitability, adequacy, and effectiveness |

From the implicit and explicit management responsibilities listed above, one can conclude that ISO 14000 expects management to be involved in virtually everything connected with the EMS. This *is* the ISO position. Management is responsible for the success of the EMS, compliance with legal and other requirements, and conformance to the EMS—regardless of whether the responsibility is explicitly or implicitly stated. Success cannot be achieved without employees at all levels willingly supporting the EMS and its objectives. Likewise, just as it is true of Total Quality Management and of ISO 9000's quality management system, an ISO 14000 EMS simply cannot be successful without complete commitment and participation of top management. Senior managers must state their commitment and demonstrate it by actively participating in the ISO 14000 effort. People follow by observing; so when employees see their leaders involved in environmental activities, they will accept that a commitment does, in fact, exist. This point, though lost on many managers, is critical for the successful implementation and operation of an EMS. ISO states the following:

> The success of the [environmental management] system depends on commitment from all levels and functions, especially from top management.[1]

## ELEMENTS OF AN EMS AND EXTENT OF APPLICATION

The components, or elements, of an EMS that conform to the requirements of ISO 14000 are as follows:

- **Management commitment**—Management commitment must be present at the start and sustained over time. If management is not committed to the objectives of ISO 14000 and actively involved in related environmental activity, no chance exists for successful integration and operation of an EMS.

- **Conforming environmental policy**—The environmental policy crafted by management, or under management's direction, is the guiding document that establishes the "overall sense of direction" and sets the environmental "principles of action"[2] for the organization.

- **Environmental planning**—For an effective EMS, the organization must identify (1) the aspects of its operation that can have environmental impact and (2) the laws, regulations, and other environmental requirements to which it must comply. Then it must plan for dealing with these items. Part of that planning must involve setting environmental objectives and targets and establishing programs to assure their achievement.

- **Organizational structure and responsibility**—The organization's hierarchy, relative to environmental aspects, and the responsibility and authority assignments for each relevant level must be clearly defined in the EMS and understood by employees.

- **Awareness and competence training**—Management is responsible for ensuring that all employees are knowledgeable about the organization's environmental aspects, policies, and commitment. It is also responsible for ensuring that employees involved with environmentally related operations are competent to execute their functions. This is accomplished through training and evaluation as established under the EMS.

- **Effective internal and external communication**—The organization must establish processes for the EMS that ensure timely, effective communication internally (with employees) and externally (with interested parties out of the organization).

- **Control of environmentally related operations and documentation**—EMS control of operations is accomplished primarily through the use of proven documented procedures for the processes that can have environmental impact and by ensuring that procedures are followed rigorously. To support this, the organization must have a documentation control system which assures that (1) proper procedures are issued for use and (2) changes follow the established approval process.

■ **Emergency preparedness and response capability**—The EMS must have procedures for emergency environmental situations. Preparedness and response capability must be achieved and demonstrated through training and practice drills which are specified in the organization's EMS.

■ **Checking/auditing and corrective and preventive action**—The EMS must translate feedback from checking, measurement, and monitoring of ongoing environmental performance into corrective and preventive actions. This is a critical part of the EMS' PDCA Cycle. Whenever a problem arises, management must find ways to correct the situation and prevent it from happening again.

■ **Recordkeeping**—The EMS must support the maintenance of important environmental records as evidence of the ongoing operation. Records will be many and varied; they will be useful to the organization and to auditors, legal and regulatory agencies, and other interested parties.

■ **Management review**—The EMS must be reviewed periodically by management for continued suitability, adequacy and effectiveness, and for continual improvement opportunities.

■ **Continual improvement**—Systems must be built into the EMS for identifying potential improvements to the EMS. Continual improvement most often occurs through the elimination of the root causes of nonconformance, but it may also be the result of incorporating entirely new processes in place of old ones, adopting new technology, or other strategies.

The elements just explained generally follow the EMS model of ISO 14001 and our adaptation presented in Figure 3-2.

The EMS should be relevant to all phases of the life cycle of the product or service, from conceptual development to ultimate disposal at the end of the product's useful life. The EMS will encompass many processes and procedures. From the standpoint of the product's life cycle, these processes and procedures come into play at different times. For example, at the very beginning of the life cycle, the design processes are most relevant. These processes should seek to minimize by design any subsequent environmental aspects of the new product. This concept can be thought of as designing for environmental integrity. Later in the life cycle the processes through which the new product is manufactured become active. Again, the developers of the manufacturing processes should seek to avoid or minimize environmental aspects during manufacture.

Ultimately the product is sold and used by the consumer. If the product is one that has environmental aspects (such as an automobile), the designers should have already done their best to minimize pollution when the product is operated properly. Beyond this, it is not the manufacturer's responsibility to try and force the consumer to use the product responsibly. Finally, the product wears out and is disposed of by the consumer. The degree to which it is recyclable was determined by the designers early in the life cycle. However, whether or not the consumer chooses to recycle the product is beyond the manufacturer's control. ISO recognizes that once a product is in the hands of a consumer, the producing organization can do little to enforce proper operation. ISO 14001 states that the EMS applies only to the extent that "the organization can control and over which it can be expected to have an influence."[3]

# EMS STRUCTURE

ISO 14000 does not attempt to suggest a structure for the environmental management system, since it would be impossible to design a structure to fit all organizations. However, ISO 14001 and 14004 specify the fundamental requirements and intents of an EMS, which must be tailored and adapted to the particular organization's operations, culture, and resources. From this the organization is expected to translate the standard's requirements and intents into an EMS structure that will support them while meeting the organization's needs. The best starting point for developing an EMS structure is a review of ISO 14000's EMS model, shown in Figure 4-1.

Each blocked element of the EMS model (such as Environmental Policy or Planning) corresponds to one of ISO 14001's five 2-digit clauses representing the requirements for the EMS.

■ Environmental Policy—Clause 4.2. The EMS is envisioned to start with the organization's commitment to prevention of pollution and the continual improvement of its environmental performance. This commitment is accompanied by a policy state-

**Figure 4-1**
ISO 14000 EMS Model

Adapted from ANSI/ISO
14001–1996, Figure 1, p. vii.

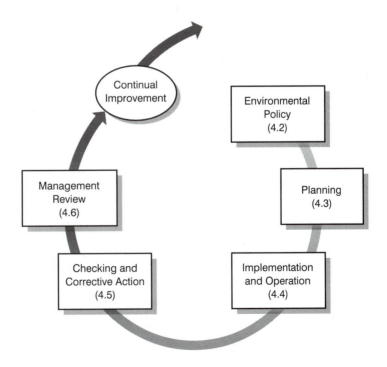

ment that outlines the philosophy and guiding principles under which the organization intends to operate its EMS. This is the first stage of the EMS structure and provides the foundation upon which the EMS is constructed and operated. The environmental policy must be revisited from time to time to assure its continued relevance and adequacy.

■ Planning—Clause 4.3. The second stage of the EMS model deals with identifying the organization's legal, regulatory, and other requirements; its significant environmental aspects; and appropriate environmental objectives and targets. It also establishes programs to assure achievement of the objectives and targets. After the initial implementation, this stage will remain active to deal with changes—in the organization's operations, activities and processes, legal and regulatory requirements, changes resulting from continuous improvement, and more. As such, it becomes the Plan stage of the closed-loop Plan-Do-Check-Adjust Cycle. (Refer to chapter 3, Figures 3-1 and 3-2, and the discussion of the PDCA Cycle.) The planning stage effectively establishes what the organization must do to achieve compliance with legal and regulatory requirements and conformance to ISO 14000 and its own environmental expectations.

■ Implementation and Operation—Clause 4.4. The model's third stage provides the tools, procedures, and resources necessary to put the EMS into sustained operation. Implementation and initial operation are achieved through the following steps:

1. Designation of responsibility and authority relative to the EMS. This may be in the form of an organization chart, job descriptions, or other suitable means of communicating to the employees.

2. Designation of the Management Representative, who will provide the day-to-day management of the implementation and ongoing functions of the EMS.

3. Providing the necessary human, technology, and financial resources.

4. Establishing and implementing the reporting policies and procedures necessary to ensure that top management is aware of EMS performance.

5. Determining training needs and providing required training for relevant personnel.

6. Establishing and implementing policies and procedures for internal and external environment-related communication.

7. Establishing and implementing policies and procedures relative to documenting the EMS.

8. Establishing and implementing policies and procedures for controlling EMS documentation.

9. Establishing and implementing policies and procedures for controlling the organization's relevant operations and processes.

10. Establishing and implementing policies and procedures for emergency preparedness and response.

The Implementation and Operation phase puts the EMS framework in place. It will require continual updating when changes, such as employee job reassignments, occur as the organization's activities or products change, as training needs change over time, and as policy and procedures evolve through continual improvement. This is the Do stage of the Plan-Do-Check-Adjust Cycle.

■ **Checking and Corrective Action—Clause 4.5.** The model's fourth stage embodies the day-to-day operation of the EMS, and deals with monitoring the performance of the EMS elements and environmental aspects and responding to nonconformances. This is accomplished through the following steps:

1. Regularly monitoring and measuring the key characteristics of the organization's activities and operations that can have a significant environmental impact.
2. Periodically evaluating compliance with relevant environmental laws and regulations.
3. Calibrating and maintaining monitoring equipment according to the organization's procedures.
4. Investigating all nonconformances.
5. Taking corrective and preventive action to mitigate environmental impacts and prevent recurrence of the same nonconformance.
6. Keeping environmental records according to the organization's procedures and policies.
7. Periodically auditing the EMS to assure continued EMS conformance to ISO 14000 and the organization's environmental plans and policy, and to determine if it has been properly implemented. (Audits may be internal or external.)

This stage will also be under constant review for updating to improve its own processes or to accommodate changes in the other stages. This stage represents Check in the Plan-Do-Check-Adjust Cycle.

■ **Management Review—Clause 4.6.** The fifth and final stage of the model deals with management reviews of the EMS. The review process requires the collection of information relevant to the EMS and the presentation of this information to senior management on a scheduled basis, with other inputs as required by the nature of the information. The purpose of the reviews is to:

1. Ensure the EMS's continued suitability.
2. Confirm its adequacy.
3. Verify its effectiveness.
4. Facilitate continual improvement of the EMS, processes, environmental equipment, and so forth.

From management's review of EMS, the processes and equipment employed, and other relevant environmental performance, it will determine what is acceptable in its present condition and what needs to be changed. This is the Adjust phase of the Plan-Do-Check-Adjust Cycle.

The structure just described may be envisioned as in Figure 4-2.

The EMS structure may also be viewed in terms of the PDCA Cycle, as in Figure 4-3.

## DOCUMENTING THE EMS

The following quote is from ISO 14001, Clause A.4.4:

> The level of detail of the documentation should be sufficient to describe the core elements of the environmental management system and their interaction and provide direction on where to obtain more detailed information on the operation of specific parts of the environmental management system.

This clause notes that there is no requirement for all documentation to be incorporated in one manual. Documents such as process procedures and work instructions are usually best kept separately due to their volume, but the environmental manual must refer to the actual location of these documents so that they can be found easily.

Four levels of EMS documentation are required; refer to Figure 4-4. As a minimum, the levels include:

- Level 1: Environmental Management System Manual. The manual must contain the organization's environmental policy, policies for each clause of ISO 14001, organization charts, emergency plans, and an index to Level 2 procedures.

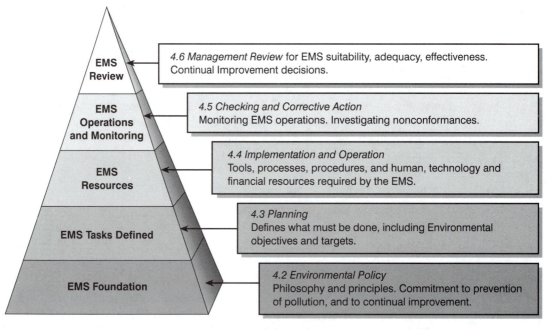

**Figure 4-2**
EMS Structure

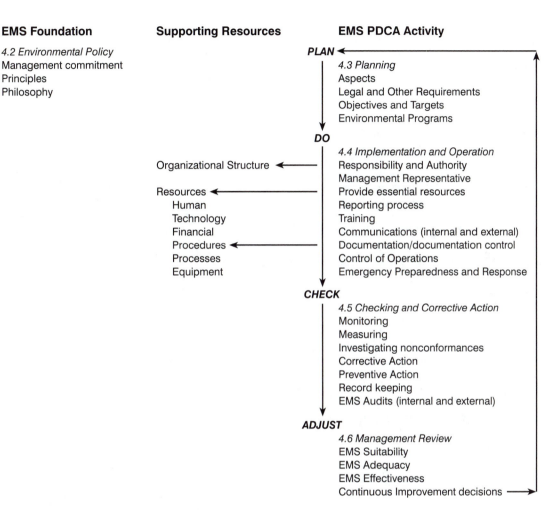

**Figure 4-3**
EMS Structure in Terms of the PDCA Cycle

- Level 2: Environmental Procedures. The procedures describe what the organization does to meet the Level 1 policies. A procedure should be included for each clause and indexed to the Level 3 practices.

- Level 3: Practices. The practices are process procedures, work instructions, and other documents by which work is accomplished.

- Level 4: Proof. These are the organization's environmental records which provide objective evidence for compliance with legal and regulatory requirements and conformance to ISO 14000 and the organization's environmental policies.

The requirements of the four levels of EMS documentation are explained in more detail in the next section.

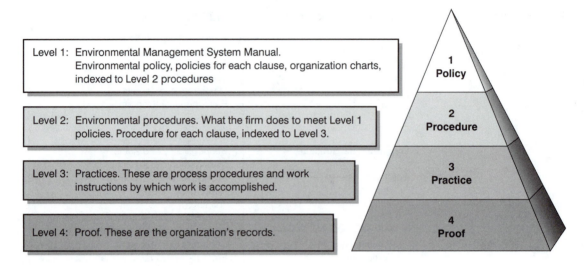

**Figure 4-4**
EMS Documentation Hierarchy

## Level 1

Level 1 is the Policy Level that, in essence, declares the organization's environmental philosophy and commitment and how it intends to conduct itself. It details the environmental issues facing the organization and plans in a general way for avoiding environmental problems as well as improving environmental performance. All of this will be in the form of the Environmental Management System Manual.

The EMS manual will include the environmental policy, with specific policies corresponding to each clause of ISO 14001, Section 4, Environmental Management System, requirements. As with ISO 9000, a good starting point for the environmental policy is a clause-by-clause duplication of ISO 14001, Section 4, substituting *will* for *shall*, *we* or *our* for *organization*, and other words as appropriate. For example, the language "The organization shall establish and maintain. . . . " in one of the Section 4 clauses can be changed to "We will establish and maintain. . . . " in the corresponding environmental policy statement.

Organization charts and other forms of documents that define the core elements of the EMS are also included in the EMS Level 1 documentation. They include information on how the documents relate and interact with each other and the organization as well as define management responsibility and authority for operating the EMS and each of its elements (Clause 4.4.1).

Although not explicitly required, the organization should include a list of the significant environmental aspects of the its activities, products, or services. Identifying them is a requirement (Clause 4.3.1), and the organization's environmental objectives and targets will evolve from them. This is the appropriate location for the list.

A section in the EMS manual should be reserved to list the current environmental objectives and targets. This can be in a list format, referring to the actual documentation of the objectives and targets, or it can be the active set of objective and target documentation. In either case regular updating will be required to account for objectives and targets being achieved and removed from the active list (Clause 4.3.3).

It is also a good idea to include the applicable legal, regulatory, and other requirements in the EMS manual, at least in list form with references to actual locations (Clause 4.3.2).

The EMS manual should include a list of environmental procedures that correspond to requirements and to environmental aspects. The procedures themselves are not located in the manual because they would make the manual unwieldy. The list should provide references to the actual location of the procedures so that anyone needing access can find them easily.

An EMS manual constructed according to the foregoing will have these elements:

- Environmental policy reflecting each of the Section 4 clauses.
- Definition of core elements of the EMS and their interrelationships.
- Clear definition of management responsibility and authority for operating the EMS and its elements.
- Current list of the significant environmental aspects.
- Current list of environmental objectives and targets (or the actual objective and target documentation).
- Copies of legal, regulatory, and other environmental requirements or a list of them with references to actual location.
- A list of the organization's procedures related to each of the Section 4 clauses, indexed to their Level 2 location.

## Level 2

Level 2, the Procedures Level, describes how the organization operates the EMS. Minimally, it should include procedures to address each procedure requirement of ISO 14001, Section 4. Clearly, only documented procedures can be included in Level 2 of the EMS documentation. One must be careful here, because while some procedures are explicitly required to be documented by the clauses, others carry only an implicit requirement that can easily be missed until the registrar indicates the error. The organization should use documented procedures " . . . to cover situations where their absence could lead to deviations from the environmental policy. . . . " (Clause 4.4.6). In most organizations, undocumented procedures will lead to all kinds of deviations, but documented procedures will lead to consistency. Properly documented procedures can be easily followed by someone new to the job, and documented procedures are the only ones that can be controlled.

Procedure requirements (explicit and implicit) are as follows:

| 4.3.1 | Procedure to identify environmental aspects |
| 4.3.2 | Procedure to identify and have access to legal and other requirements |
| 4.4.2 | Procedures to make employees aware of [various environmental issues] |
| 4.4.3 | Procedures for internal communications |
| 4.4.3 | Procedures for receiving, documenting, and responding to communications from external interested parties |
| 4.4.5 | Procedures for controlling documents |
| 4.4.5 | Procedures for creating and modifying documents |
| 4.4.6 a) | Procedures for activities where absence could lead to deviations |
| 4.4.6 b) | Procedures stipulating operating criteria |
| 4.4.6 c) | Procedures related to environmental aspects of goods and services used by the organization |
| 4.4.6 c) | Procedures for communicating relevant procedures and requirements to suppliers and contractors |
| 4.4.7 | Procedures for emergency preparedness and response |
| 4.5.1 | Procedures for monitoring and measuring key operational characteristics |
| 4.5.1 | Procedure for periodically evaluating compliance |
| 4.5.2 | Procedure to define responsibility and authority for handling and investigating nonconformance |
| 4.5.3 | Procedures for identification, maintenance, and disposition of environmental records |
| 4.5.4 | Procedure for periodic EMS audits |

In addition to these requirements, the organization may use several other environmentally related procedures, which also should be included in the Level 2 documentation.

The procedures together describe step-by-step what the organization does to meet Level 1 policies. Ideally, there should be procedures related to each of the ISO 14001 Section 4 clauses. The Level 1 list of procedures becomes the table of contents for Level 2. In turn, Level 2 should contain a list of the process procedures (practices) contained in Level 3.

## Level 3

In Level 3, the Practices Level, the process procedures (i.e., the actual work instructions for activities relevant to the EMS) reside. They represent what the employees do in their operational activities. These process procedures, or work instructions, will provide detailed step-by-step instructions dealing with significant environmental aspects of the organization's activities or for activities required by the Standard. Organizations involved in Total Quality Management or ISO 9000 will probably already have these process procedures documented; others may not. Organizations that have not documented their process procedures should do so for consistency of operations. The Level 2 list of process procedures (practices) becomes the table of contents for Level 3. In turn, Level 3 should list the Level 4 forms and records associated with the practices and their location.

## Level 4

Level 4, Proof Level, is the repository for all forms, records, and the like which represent the objective evidence (proof) that the EMS is, or is not, functioning as it should. In accordance with ISO 14001, these would include, as a minimum, EMS records such as the following:

- Record of decision concerning external communication (Clause 4.4.3)
- Documentation of communication from external interested parties (Clause 4.4.3b)
- Records related to monitoring and measurement (Clause 4.5.1)
- Records related to maintenance and calibration of monitoring equipment (Clause 4.5.1)
- Records of changes to documented procedures resulting from corrective or preventive action (Clause 4.5.2)
- Training records (Clause 4.5.3)
- Records of results of audits and reviews (Clause 4.5.3)

These records will be useful to the organization, and they will be the focus for registrar audits, serving as objective evidence that the organization is or is not in conformance with the ISO 14000 Standard.

Both the registrar and the organization itself also have vested interest in compliance with legal, regulatory, and other environmental requirements. Therefore, it is obvious that in addition to the conformance records listed above, compliance records associated with legal, regulatory, and other requirements should also be maintained as Level 4 documents.

Documentation of the EMS for ISO 14000 will require a significant effort for organizations that are not already operating under the principles of Total Quality Management, ISO 9000, or government requirements that demand documented processes, procedures, and other similar criteria. ISO 14000 documentation is intended to assist the organization, not to burden it. Procedures should be kept as simple as possible and be consistent with requirements and good judgment. The auditors will require that the organization does exactly as it says it will do in its procedures. Consequently, the best course is to begin simply and add detail later if necessary. Do not make documentation unnecessarily complex.

## VERIFYING THE INTEGRITY OF THE EMS

For registered organizations, two levels of audits are used to verify the integrity of the EMS: external audits performed by the registrar and internal audits performed by the organization's employees. Registrars will normally perform periodic surveillance audits at least annually. The registration audit, or "conformity assessment," and surveillance audits are explained in chapter 6.

Clause 4.5.4 establishes the requirement for periodic internal audits. These audits determine (1) if the EMS conforms to ISO 14000 and the requirements of the environmental policy and plans and (2) if the EMS has been properly implemented and utilized. In reference to the second item, the object is to determine whether the EMS has been deployed throughout the organization and put into practice and whether the policies, practices, and procedures which comprise the EMS are being rigorously followed.

Internal EMS audits are usually performed by employees of the organization, although the organization can employ outside auditors to perform its internal audits. The two areas of concern when using employees for internal audits are as follows:

- First, the employees must be able to conduct the audit objectively and impartially. This means that internal auditors must be independent of the function being audited. For example, an employee of the finance department could audit the human resources department, but not the finance department.

- Second, the employees selected as internal auditors usually will not be experienced auditors. To achieve maximum benefit from internal audits, the auditors must be trained in auditing techniques, practices, and psychology.

As explained previously, the organization must have an EMS audit program that is supported by procedures. The program and procedures should cover the following:[4]

1. *The organization's activities (functions, departments) to be considered in audits.* All environmentally important activities (i.e., those associated with aspects which may have environmental impacts) will be included.

2. *The frequency of internal audits.* The frequency with which an activity is audited is to be a function of the activity's environmental importance, and past audit results and performance.

3. *The responsibilities associated with managing and conducting audits.* Roles, responsibilities, and activities of auditors and lead auditors must be clearly established in the procedure. (See ISO 14011, Clause 4.2.)

4. *The communication of audit results.* Audit results must be communicated to management in accordance with the procedure and the audit plan. ISO 14000 does not specify the method of communicating, but it does mention the audit report. Clearly, a written report is required. In addition, we suggest that the report be accompanied by an oral presentation, during which management can obtain any needed clarification from the audit team. (See ISO 14011, Clause 5.4.)

5. *Auditor competence requirements.* Employees carrying out the internal audits should be properly trained.[5] The guidance for auditor qualification criteria is found in ISO 14012 and will be considered daunting by most organizations. At first glance one would assume that ISO 14012 was aimed at external auditors employed by registrar firms. However, Clause 1: Scope states, " . . . and is applicable to both internal and external auditors." Fortunately this is only guidance, because in our experience few organizations could meet the education, experience, and training qualifications specified.

6. *Audit methods.* Internal auditors will collect the information they require from the environmental records maintained by the organization, through interviews with relevant personnel, and through observation of operations. The information gathered will be audit evidence which will be compared with the audit criteria to develop conclusions. The conclusions will be conveyed to management via a written audit report, perhaps accompanied by an oral presentation to promote understanding.

Considerations for EMS audits include the following.[6] (The reference document, ISO 14010, Guidelines for Environmental Auditing—General Principles, is generic in nature. In this section internal audits are discussed; therefore, the client will normally be management.)

- **Requirements for the audit.** First, the audit should focus on clearly defined and documented subject matter. Second, the parties responsible for that subject matter must be identified, their names documented, and available for the auditors.

- **Objectives and scope of the audit.** The client, usually the organization's management, defines the objectives of the audit. The scope of the audit (boundaries and extent) is determined by the lead auditor, consulting with the client, to meet the audit objectives. The objectives and scope should be communicated to personnel involved in the activities to be audited.

- **Objectivity, independence, and competence.** Members of the audit team should be independent of the activities they audit, and they should be objective and without bias or conflict of interest. Audit team members should have the knowledge, skills, and experience to perform audits of assigned activities.

- **Due professional care.** Auditors must exercise the care, diligence, skill, and judgment expected of any auditor. Of utmost importance is confidentiality and discretion between auditor and persons interviewed and client (management). There should be no attribution (e.g., "John said. . . . "). No information obtained during the audit should be disclosed to any third party or anyone outside the organization, unless required by law.

- **Systematic procedures.** To ensure reliability and consistency, environmental audits should be conducted in accordance with these principles and the guidelines given in ISO 14011.

- **Audit criteria, evidence, and findings.** Audit criteria normally are the policies, practices, procedures, and requirements against which auditors compare the evidence collected during the audit. The audit criteria for a particular audit should be selected by the client and the lead auditor and then communicated to the auditee prior to the audit. Relevant information should be collected during the audit, and it should be analyzed, interpreted, and recorded as audit evidence in order to determine whether the audit criteria are met. Audit evidence should be of such quality and quantity (reliable, and, ideally, from multiple sources) that multiple auditors working independently would reach the same conclusions.

- **Reliability of audit findings and conclusions.** Since environmental audits take place in a discreet, short period of time, any evidence gathered can be no more than a sample. The audit process should be designed to give the auditors and management confidence in the reliability of the audit findings and conclusions. It is necessary that auditors recognize the uncertainty of audit findings and conclusions and take this into account in planning and conducting the audit.

- **Audit Report.** A written audit report should be submitted to management (internal audits). The audit report may, at the discretion of the lead auditor and the client, include the following:

    1. Identification of the activity or activities audited.
    2. Audit objectives and scope.
    3. Audit criteria.
    4. Period covered by the audit and dates of the audit.
    5. Identification of audit team members.
    6. Identification of auditee participants.
    7. Statement of confidentiality.
    8. Distribution list for the audit report.
    9. Summary of the audit process and any obstacles encountered.
    10. Audit conclusions.

Note that the audit report does not normally include any corrective actions required. It is usually up to the organization or relevant activity to determine what corrective action should be taken. However, recommendations for corrective action may be provided if there has been prior agreement to do so.

## REVIEW AND EVALUATION OF THE EMS

ISO 14001, Clause 4.6, requires that management periodically review the EMS for continued suitability, adequacy, and effectiveness; further, it must review the EMS to maintain continual improvement. One might question the necessity for these reviews once the EMS is in place and accomplishing its objectives. However, nothing remains static for long. The organization's products or services will change over time to meet the

---

### ISO 14000 INFO

*The primary technique in dealing with audit uncertainty is finding corroborating evidence. A single piece of evidence carries a low reliability factor, but when supporting evidence turns up one, two, or more times, the reliability increases correspondingly.*
    *Goetsch and Davis*

changing demands of customers. Those changes will result in changes to processes and procedures and may eliminate some environmental impacts while creating new ones. The organization will find ways to improve its products, processes, and procedures, resulting in the need for new EMS documentation. Environmental technology will also change, perhaps making it possible to improve environmental performance by adopting the new technology. The legal and regulatory agencies will add new environmental requirements or modify existing ones. With these and many other changes taking place, the need for periodic management reviews of the EMS is critical.

ISO 14000 does not presume to establish a review schedule or even suggest one. ISO 14001, Clause 4.6, states, " . . . at intervals that [management] determines." Clause A.6 uses the phrase *defined intervals*. ISO 14004, Clause 4.5.2, uses *appropriate intervals*. Management has to determine an appropriate interval and define it in a schedule that is understood by the relevant employees. Although ISO 14001, Clause A.6, notes that reviews should be comprehensive, it also states that not all elements of the EMS need to be reviewed at once. Therefore, a complete and comprehensive review of the EMS could occur over the course of several scheduled mini-reviews. A monthly interval seems to be appropriate. The total EMS should be discussed, with emphasis on the more environmentally critical elements. Once a review interval is established, management can evaluate its effectiveness and alter the schedule if necessary. Under no circumstances would a review cycle of longer than three months be adequate; an interval less frequent than quarterly will send the message that reviews, and perhaps environmental performance, are unimportant, in which case the focus on environmental issues may be lost.

EMS reviews should include the following:

■ Overall environmental performance measured against the EMS.

■ Review of progress or problems relative to the achievement of all environmental objectives and targets.

■ Findings from audits and corrective or preventive actions planned or taken.

■ Evaluation of EMS effectiveness.

■ Evaluation of continued suitability of the EMS in light of[7]
   • changing legislation
   • changing expectations and requirements of interested parties
   • changes in the organization's products or activities
   • advances in science and technology
   • lessons learned from environmental incidents (inside or outside the organization)
   • market preferences
   • reporting and communications

The intent of the reviews is that the EMS and all of its components will be continually improved to ensure that environmental performance is also continually improved. Therefore, continual improvement is to be a significant element of the management reviews. The management review process and procedures should ensure that the continual improvement process is always a major element of the reviews. The continual improvement process includes the following steps:[8]

■ Identify areas of opportunity for EMS improvement. Remember, the EMS includes processes, procedures, and work instructions.

■ Determine root causes of nonconformances and deficiencies. ISO 14000 is not looking for workaround solutions or "Band-Aids," although such quick fixes sometimes may be required to mitigate an environmental incident. ISO 14000 seeks elimination of the root cause responsible for the nonconformance. A quick fix may minimize an incident or get the process running again, but real improvement is possible only by eliminating root causes.

■ Develop and implement plans for action to eliminate root causes.

■ Verify the effectiveness of the actions.

■ Document any changes in procedures resulting from improvement actions.

■ Make comparisons with objectives and targets.

Since ISO 14001, Clause 4.6, requires EMS reviews to be documented, all observations, conclusions, and recommendations must be documented and retained as records.

## SUMMARY

1. The EMS is the part of the overall management system that includes the organizational structure; planning; assignment of responsibilities; processes, procedures, and practices; resources; and all other elements necessary to achieve, maintain, and continually improve sound environmental performance.

2. Commitment to ISO 14000 by top management is crucial; the EMS will fail without it. Management is responsible for establishing and maintaining the objectives and targets, programs, procedures, and documents required by ISO 14001, Clause 4. Management is also responsible for defining and communicating roles and responsibilities and providing training and resources. Top management is responsible for developing, documenting, communicating, and implementing the environmental policy.

3. The necessary elements of an EMS are as follows: management commitment, conforming environmental policy, environmental planning, organizational structure and responsibilities, awareness and competence training, internal and external communication, control of environmentally related operations and documentation, emergency preparedness and response capability, checking (auditing and corrective/preventive action), recordkeeping, management review, and continual improvement.

4. The structure of the ISO 14000 EMS model is as follows: environmental policy, planning, implementation and operation, checking and corrective action, and management review.

5. It is important to document the EMS completely. The level of detail should be sufficient to describe completely all the core elements of the EMS and how they interact.

It should also provide references to locations of more detailed information about the operation of specific elements.

6. The integrity of the EMS is verified at two levels: internal audits and external audits. Internal audits are performed by employees or consultants brought in by the organization; external audits are conducted by the registrar. When an organization decides to use its own employees to conduct internal audits, it should take care to satisfy two concerns. The first is, Use employees who are independent of the functions being audited to ensure objectivity. The second is, Make certain employees receive the necessary training.

7. Periodic reviews of the EMS should include the following: the organization's overall environmental performance measured against the EMS; review of environmental objectives and targets, including notation of progress made and problems confronted; findings from audits and corrective/preventive action taken or planned; evaluation of the effectiveness of the EMS; and evaluation of the continued suitability of the EMS.

## KEY CONCEPTS

Awareness and competence training

Checking and auditing and corrective and preventive actions

Conforming environmental policy

Continual improvement

Control of environmentally related operations and documentation

Effective internal and external communication

Emergency preparedness and response capability

EMS

Environmental planning

Management commitment

Management review

Organizational structure and responsibility

Recordkeeping

## REVIEW QUESTIONS

1. What is the purpose of the EMS?
2. Explain the relationship between policy and management systems.
3. Describe or explain the following EMS concepts:
   - Management commitment
   - Conforming environmental policy
   - Environmental planning
   - Organizational structure and responsibility
   - Awareness and competence
   - Effective internal and external communication
   - Recordkeeping
   - Continual improvement

4. Draw a diagram that illustrates the structure of the ISO 14000 EMS.

5. Explain each major component in the EMS model drawn in Question 4.

6. Describe the level of detail required when documenting the EMS.

7. Explain the two levels (types of audits) used to verify the integrity of the EMS.

8. The organization's audit program should have procedures that cover at least six areas of concern. List the six areas.

9. Explain how the following concepts apply to an organization's internal audit program: objectivity, independence, and competence.

10. An audit report should contain what elements?

11. List the main elements of an EMS review.

12. What are the steps that lead to continual improvement?

## CRITICAL-THINKING PROBLEMS

The following activities may be assigned as individual, group, or discussion activities to be completed in class or out of class:

1. Research the difference between policy and procedures. Write a brief essay that explains your findings, and give at least one illustrative example.

2. Explain in your own words the practical aspects of why the commitment of top management is so important to the success of any major undertaking in an organization. Locate an example of an effort that failed in business, government, or the military because it lacked top-level commitment.

3. If you were the CEO of a company and were committed to gaining ISO 14000 registration status, how would you go about securing organizational commitment from all employees at all levels?

4. An organization's EMS should be relevant to all phases of a product or service's life cycle. This is what makes the concept of "designing for environmental integrity" relevant to ISO 14000 registration. Develop an example that illustrates how this concept might be applied.

5. In applying the ISO 14000 EMS model in Figure 4-1, it is important to ensure that the organization's environmental objectives and targets are being met, as are all applicable local, state, and federal regulations. Yet the model requires continual improvement, which implies performance that exceeds objectives, targets, and regulations. Explain the rationale for continual improvement even after applicable standards are being met.

## DISCUSSION CASE

### A Tale of Two Companies

A. Brown Processing (ABP) and Exotic Meats, Inc. (EMI), are the leading processors of nontraditional meat products in their region. They process ostrich, emu, deer, buffalo, elk, and other so-called exotic animals for specialty restaurants. Each company is the other's principal competitor. Consequently, they both are always looking for strategies that will give them a competitive advantage.

The CEO of ABP thinks ISO 14000 registration will show the community that his company is a responsible corporate citizen and, in the long run, will decrease his operating costs by eliminating expensive litigation and fines from regulatory agencies. He is committed to pursuing ISO 14000 registration. Unfortunately, his executive managers and employees do not share his commitment.

EMI has the opposite problem. The CEO's executives, managers, and employees have encouraged him to consider ISO 14000 registration, but he thinks it is a waste of time.

## Discussion Questions

Discuss the following questions in class or out of class with your fellow students:

1. What types of barriers must the CEO of ABP overcome in order to pursue ISO 14000 registration, and how might he do so?

2. How can the executives, managers, and employees of EMI convince their CEO to commit to ISO 14000 registration?

3. In which company would you rather be the Environmental Manager? Why?

## ENDNOTES

1. ANSI/ISO 14001-1996, Introduction, p. vi.

2. ANSI/ISO 14004-1996, Clause 4.1.4, p. 6.

3. ANSI/ISO 14001-1996, Clause 1, p. 1.

4. Ibid., Clause A.5.4, p. 9.

5. ANSI/ISO 14004-1996, Clause 4.4.5, p. 22.

6. ANSI/ISO 14010-1996, Clauses 3 and 4, pp. 2-3.

7. ANSI/ISO 14004-1996, Clause 4.5.2, p. 23.

8. Ibid., Clause 4.5.3, p. 24.

# EMS Documentation

## DEFINITION OF DOCUMENTATION

*Document* in the ISO 14000 context is something written or printed that is relied upon to record or prove something. *Document* also is used as a verb to describe the act of recording information. *Documentation* refers to the documents or references held for records and proof. For ISO 14000 purposes **documentation** includes written and pictorial information used for defining, specifying, describing, or recording the activities and processes of the organization, especially those related to environmental performance. This documentation may be in hard copy or electronic format. ISO 14000 documentation may include (but is not limited to) the following elements:

- Written policies
- Written procedures
- Written work instructions
- Written plans
- Written records

- Formal drawings
- Sketches
- Legal requirements
- Regulatory requirements
- Other requirements

## ISO 14000 DOCUMENTATION REQUIREMENTS

As explained in chapter 4, EMS documentation has four levels. The first is the Policy Level, serving as the foundation of the EMS. Second is the Procedures Level, establishing what the organization does in order to satisfy the first-level policies. Third is the Practices Level, providing **work instructions** for the relevant employees. Fourth is the Proof Level, the repository for forms and records related to the EMS. The four levels are illustrated in Figure 5-1.

In this section documentation explicitly required by ISO 14001 is listed, as well as documentation for which a requirement is inferred. Whether a requirement is explicit or implicit is determined by the wording of the Standard. For example, Clause 4.2 e) explic-

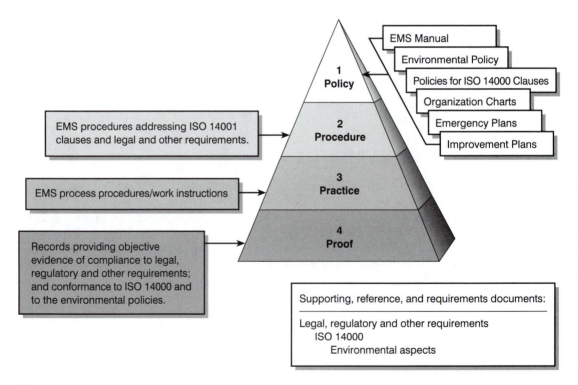

**Figure 5-1**
ISO 14000 Documentation System

itly requires that the organization's environmental **policy** be documented. On the other hand, Clause 4.3.1 requires the organization to establish and maintain procedures to identify and list its environmental aspects and keep them up-to-date, but there is no stated (explicit) requirement for these procedures to be documented. The procedures are to be used continually and forever, so it is practical to document the procedures. This, then, appears to be an implicit requirement for documentation.

Clause 4.4.6 adds another dimension to its requirements for documenting procedures related to operations that have environmental aspects. Subclause a) requires that such procedures be documented when the absence of documentation " . . . could lead to deviations from the environmental policy and the objectives and targets." With the possible exception of a one-person organization, the absence of documented procedures will inevitably lead to deviations, thereby implicitly requiring that all procedures be documented.

In the following list of documentation requirements, *explicit requirements are in italics*. To conform with ISO 14001, the organization must adhere to these requirements for documenting procedures, plans, and other EMS elements. Implicit documentation requirements are listed in regular print, followed by the rationale for listing them as implicit requirements.

| Document Required | Requiring Clause |
|---|---|
| *Environmental policy* | 4.2 e) |
| Environmental aspects identification procedure (This clause requires the development and use of a procedure, but does not explicitly require its documentation. However, because the subject is too complicated for an unwritten procedure, it should be documented.) | 4.3.1 |
| Legal and other requirements identification-and-access procedure (A procedure for identifying and providing access to legal and other requirements will be too complicated to leave undocumented.) | 4.3.2 |
| *Environmental objectives and targets* | 4.3.3 |
| Programs for achieving objectives and targets (This is too complicated to leave undocumented.) | 4.3.4 |
| *Designation of responsibility for achieving objectives and targets* | 4.3.4 a) with a link from 4.4.1 |
| *Roles, responsibility, and authority defined, communicated* | 4.4.1 |
| Employee awareness procedures (The scope of environmental awareness training and consistency of application make documentation of procedures necessary.) | 4.4.2 |
| *Records related to training* | 4.4.2 with a link from 4.5.3 |

| | |
|---|---|
| Internal communications procedure (Procedures must be understood and consistently applied, which requires documentation.) | 4.4.3 |
| External communications procedure (Procedures must be understood and consistently applied, which requires documentation.) | 4.4.3 |
| *External communications documentation* | 4.4.3 b) |
| *Core elements of the EMS and their interaction (e.g., EMS Manual)* | 4.4.4 a) |
| *Direction to related documentation (i.e., cross-referencing, location)* | 4.4.4 b) |
| Document control procedures (The only way that a document control system will work is through a set of documented procedures designed for that purpose.) | 4.4.5 |
| *Document change approval authority* | 4.4.5 b) with a link from 4.4.1 |
| *Document creation and modification responsibility* | 4.4.5 with a link from 4.4.1 |
| *Operational control procedures (where their absence could lead to deviations from the environmental policy, objectives, and targets)* | 4.4.6 |

Note: Clause 4.4.6 should be interpreted as applying to all operational procedures (process procedures, work instructions) as well as procedures required by ISO 14001 that are related to significant environmental aspects.

| | |
|---|---|
| Procedures for purchased goods and services having significant environmental aspects (If these procedures are not documented, employees engaged in procuring goods and services may not consistently communicate relevant procedures and requirements to suppliers.) | 4.4.6 c) |
| Emergency preparedness and response procedures (Although Clause 4.4.7 does not explicitly require documentation of these procedures, this is clearly the clause's intent. It is unlikely that the clause could be satisfied in the absence of documented procedures.) | 4.4.7 |
| *Key operational characteristics monitoring and measuring procedures* | 4.5.1 |
| *Monitoring and measuring records* | 4.5.1 |
| *Monitoring and measuring equipment calibration and maintenance records* | 4.5.1 |

| | |
|---|---|
| *Legal and regulatory compliance evaluation procedures* | 4.5.1 |
| Nonconformance and corrective and preventive action procedures (These procedures must address (1) responsibility and authority for handling nonconformances, (2) taking action on non-conformances, (3) initiating and implementing corrective action, and (4) updating procedures affected by corrective or preventive actions. In a practical sense, there is no other way to ensure that all this is done, except through documented procedures.) | 4.5.2 |
| *Records of changes to procedures resulting from corrective or preventive actions* | 4.5.2 |
| Procedures for identification, maintenance, and disposition of environmental records (While Clause 4.5.3 does not explicitly require these procedures to be documented, their nature and the detail required make the requirement implicit.) | 4.5.3 |
| *EMS audit results records* | 4.5.3 |
| Procedures for an EMS audit program (Such a procedure is too complex to be left undocumented.) | 4.5.4 |
| *EMS management review records* | 4.6 |

As we noted in our discussion of Clause 4.4.4, EMS Documentation, of all the specific ISO 14001 procedures that are required, few are explicitly required to be documented. Common sense, along with the caution of Clause 4.4.6, which requires that documented procedures be used "where their absence could lead to deviations from the environmental policy and the objectives and targets," suggests that all procedures related to ISO 14000 be documented.

Beyond the procedures, all other elements of the EMS documentation are straightforward. **Policies** must be documented. **Practices** (i.e., the work instructions) also must be documented if the absence of documentation could lead to deviation. Finally, **proof**, or the records, must be documented.

---

### ISO 14000 INFO

*Regarding the need for documented procedures, Joseph Cascio says, "Word-of-mouth information rarely is communicated consistently. Only written information—clearly written—is constant."*[1]

ISO 14000 requires that environmental policies be understood and that **procedures** be consistent with ISO 14001 requirements and the organization's environmental policy. More implicit requirements for EMS documentation exist, both from ISO 14000 and arising from the nature of environmental protection.

■ **Clarity.** ISO 14001 requires that documentation "be sufficiently clear to be capable of being understood by internal and external interested parties. . . . "[2] This reference applies to the environmental policy but is implicitly applicable to all EMS documentation. Users of the documentation may find comprehension difficult if it uses jargon and acronyms that are unfamiliar to employees and outside interested parties or if it uses a writing style that is not attuned to them. When it is necessary to use jargon or acronyms, definitions must be provided. Documentation should be written in the simplest terms possible consistent with the subject matter. Auditors will seek to confirm that employees understand the documentation. If they find that employees do not understand the relevant documentation, the documentation, not the users of the documentation, will be declared deficient.

■ **Effectiveness.** " . . . the primary focus of organizations [concerning EMS documentation] should be on the effective implementation of the environmental management system and on environmental performance."[3] ISO 14001 implicitly requires that EMS documentation be **effective**, although one does not find the word used with every document requirement. This means that procedures implemented in order to consistently produce some quantitative or qualitative result must, in fact, produce that result. Procedures which do not consistently produce the desired results are unacceptable. Beyond ISO 14000 conformance considerations, ineffective procedures could have serious legal implications where they result in failure to comply with legal or regulatory requirements.

## FORMAT REQUIREMENTS

ISO 14001 does not specify the form or format for the various documents and records required. It generally describes what must be documented but allows the individual organization to determine the format. This is appropriate given the fact that ISO 14000 organizations range from small businesses employing just a few people to huge multinational corporations. Simple, low-cost, low-maintenance documentation may suffice for a small, simply structured business, but it will be inadequate for the more numerous and complex functions of a larger organization. The key is that the documentation should not be more elaborate than necessary; the simplest documentation that meets the organization's needs and, at the same time, is responsive to ISO 14001 and legal and regulatory requirements is best.

Although ISO 14000 does not dictate a particular format, the EMS documentation must use appropriate, consistent formatting for its policies, procedures, process work instructions, reports, drawings, charts, emergency plans, and other documentation. A number of elements should always be part of the written documentation for policies, procedures, and work instructions, as follows:

| Element | Purpose |
|---------|---------|
| Organization name or logo | Identify organizational ownership |
| Title | Name the document |
| Related To | Cross-reference ISO 14001 clause or other requirement |
| Document Number | Identify and control document |
| Revision Status | Identify and control revisions of document |
| Date of Issue/Effectiveness | Support document control and applicability |
| Review/Approval Signatures | Signify authority and control |
| Purpose | Describe intent of document |
| Body | Impart information or content of document |

If the organization already has a documentation system under ISO 9000 or Total Quality Management, the formats used may suffice. If the organization is creating an initial documentation system, it must develop formats for each document type (drawing, policy, procedure, report, work instruction) that standardizes terminology, logos, provision for revision number and date, approval authority, and the other elements just listed. To be most effective, a procedure should be immediately identifiable as a current, authorized procedure; a policy, as a current, authorized policy; a work instruction, as a current, authorized work instruction; and so on.

Several commercial software packages are available that are designed to help organizations develop documentation that is acceptable to ISO. One advantage of these packages is the format discipline imposed by the software. If several people are involved in writing the documentation, the documents still will have a common format that promotes identification, understanding, and document control.

An example of a suitable format for a procedure was shown in Figure 3-8. For a typical environmental objective and target action plan format, see Figures 3-4 and 3-5. It is important to remember that while the formats chosen must promote identification, understanding, and document control, they must also work for the organization. It is possible to become so fascinated with a new documentation system that the reason for the documentation is lost. The objective is to help the organization implement and maintain an EMS that consistently results in environmental performance required by the organization's environmental policy, as well as applicable legal, regulatory, and other requirements. Try to find the simplest documentation system that will accomplish this objective.

## REFORMATTING PREEXISTING DOCUMENTATION

If the organization already has most of the required documentation, but in a format incompatible with ISO 14000 requirements, the task is to reformat the documents. This usually means copying the text to the proper formats. Since the writing is already done, this is usually simple, although time-consuming. The computer programs mentioned earlier might make the task easier. Keep in mind that the format of new documentation must promote identification, understanding, and document control, and it must work for the organization.

## COMBINING REQUIRED DOCUMENTATION

**Combining documentation** is a task that is separate from formatting or **reformatting**. It can be illustrated best by example. Consider ISO 14001, Clause 4.4.5, Document Control. The organization might develop (or have) at least nine separate procedures to satisfy this clause, as follows:

1. Procedure for locating documents (4.4.5 a)
2. Procedure for periodic review of documents (4.4.5 b)
3. Procedure for distribution of current documents (4.4.5 c)
4. Procedure for removal of obsolete documents (4.4.5 d)
5. Procedure for retention of documents for historical and legal purposes (4.4.5 e)
6. Procedure for document legibility (4.4.5)
7. Procedure for document formats (4.4.5)
8. Procedure for document maintenance and retention (4.4.5)
9. Procedure for document creation and modification (4.4.5)

The organization could handle document control by applying the nine procedures, or it could decide to have a single document control procedure that covers all nine areas. Additionally, closely related elements of the nine procedures could be combined, resulting in perhaps three or four procedures. For example, the organization might combine the sixth and seventh procedures (legibility and formats) into one, or it could combine the third, fourth, and ninth procedures (distribution, removal, and creation and modification). Similar applications can be found throughout the ISO 14001 clauses.

The organization may even combine procedures that have their corresponding requirements in different clauses. For example, Clause 4.3.3, Objectives and Targets, and Clause 4.3.4, Environmental Management Program, offer such an opportunity since the environmental management programs are directly related to objectives and targets.

It may be possible to combine documents across different systems. For example, documentation could be developed for use with both the ISO 14000 environmental management system and ISO 9000 quality management system, or even other systems used by the organization.

The individual organization must determine the best approach—a large number of narrowly focused documents or a smaller number of broadly focused documents—depending on its unique operations, environment, and culture. Items to consider in deciding whether to combine documentation or in determining the degree of combining include the following:

■ How will the documents work best for the people who have to use them?

■ Will combining documentation increase or decrease the organization's workload in:
- establishing the documents?
- using the documents?

- controlling the documents?
- updating the documents?

■ Which philosophy is likely to cost less in the long run?

The conclusions of one organization may be different from those of another that has different circumstances. Therefore, organizations should review each documentation element individually, and then combine them or keep them separate based on the necessities of the individual situation. Organizations should exercise caution, however, because too much combining can create problems when it comes to keeping documents current and using them. Carried too far, combining can make the system less efficient, especially for users of the documents.

## CROSS-REFERENCING REQUIRED DOCUMENTATION

Keep in mind that ISO 14000 allows organizations to refer to some documentation instead of including it in particular documents. For example, an EMS manual may refer to an organization's environmental procedures rather than actually including them in the manual itself. Specific process procedures (work instructions) may be referenced in the same way within the EMS manual or within the related procedures. Reference to legal, regulatory, and other requirements is also essential. Supporting documentation such as the environmental aspect list may also be included in the EMS manual by reference.

**Cross-referencing** should occur in two directions. For instance, the EMS manual can contain a reference to a particular procedure and its location; similarly, the procedure itself can contain a reference (page, paragraph, clause of the EMS manual) and the ISO 14001 clause to which it is directed. Most ISO 14000 organizations use cross-referencing in their documentation systems to preserve order and simplicity.

The four levels of documentation required by ISO 14000 may rely on external supporting documentation and elements of successive levels in order to attain or verify their intended function. Cross-referencing between the levels and supporting documentation is necessary if the EMS is to function efficiently; refer to Figure 5-2.

## STRUCTURE OF THE DOCUMENTATION SYSTEM

The structure of the ISO 14000 documentation system was discussed in chapter 4 in the section Documenting the EMS. It was discussed in greater detail in chapter 3 in the section dealing with ISO 14001, Clause 4.4.4, Environmental Management System Documentation. The structure of the ISO 14000 documentation system is often depicted as a pyramid as shown in Figures 3-6, 5-1, and 5-2. This is the typical illustration for the ISO 14000 documentation system, but not the only one.

The organization should design its system for compatibility with its environmental aspects, its everyday needs, with ISO 14001, legal and regulatory requirements, and other environmental requirements to which it subscribes. The organization must have an environmental policy. It also must have procedures addressing the requirements of

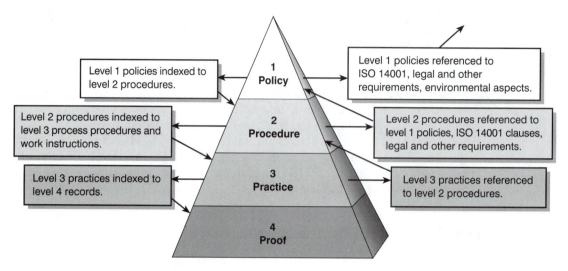

**Figure 5-2**
EMS Documentation Cross-Referencing

ISO 14001 and legal and regulatory requirements. Further, it must keep environmental records. However, beyond the guidance found in ISO 14001 and 14004, organizations have a great deal of latitude in organizing their documentation systems.

ISO 14000 even will allow EMS **documentation** to be integrated with other documentation systems. For example, such systems may be integrated with an ISO 9000 system or another documentation system already employed. Regardless of how the organization structures its EMS documentation, it is important to remember two things: (1) Users must know what applies to their operations, and (2) it must be easy for employees and auditors to locate the relevant document and its supporting documentation.

## ELECTRONIC DOCUMENTATION

ISO 14001 (Clause 4.4.4) states that EMS documentation may be in paper or electronic form. ISO 14004 (Clause 4.3.3.2) states that documents can be in any medium. Organi-

---

| ISO 14000 INFO |
| --- |
| *"This [EMS] documentation may be integrated with documentation of other systems implemented by the organization. It does not have to be in the form of a single manual."* <br> *ISO 14001, clause A.4.5* |

zations are free to choose from the available media (paper, electronic, film), although paper and electronic media appear to be the most viable approaches.

An all-electronic documentation system has advantages. For example, an **electronic documentation** system will make it easier to ensure that documentation is kept up-to-date and that any document used is the latest approved version. Distribution of new releases is simplified; and private collections of documents, which lead to the use of superseded documents, are eliminated. A disadvantage of electronic documentation is availability. Documentation such as work procedures and work instructions must be readily available to employees who need it, at their work locations. If an employee does not have the equipment necessary to view the documentation at the work location, the procedure is not readily available.

From a practical standpoint, most organizations with electronic documentation systems find it necessary to print copies of the documents for day-to-day use. Even so, the advantages of usefulness, speedy revising, orderliness, and compactness make the idea of an electronic master file worth exploring.

## SUMMARY

1. Documentation refers to the documents or references held for records or proof, including written and pictorial information used for defining, specifying, describing, or recording the activities and processes of the organization, especially those related to environmental performance.

2. ISO 14000 requires that all four levels of the EMS be appropriately documented. These levels are as follows: (1) Policy, (2) Procedures, (3) Practices, and (4) Proof. Typical documents include the EMS manual, environmental policy; policies for ISO 14001, Section 4, clauses; process procedures; work instructions; organizational charts; emergency plans; and improvement plans.

3. ISO 14001 specifies neither the form nor the format for the various documents and records required. However, it is expected that all documentation will use appropriate and consistent formatting.

4. If appropriate documentation already exists, it may be reformatted (as necessary) and used to fulfill the requirements of ISO 14001. When reformatting documentation, the organization should pay close attention to the issues of consistency.

5. Combining procedures in a documentation package is allowable. Several options for combining documentation are as follows: (1) one document that covers more than one subject, (2) combining a large number of procedures into a smaller number (e.g., combining ten into four), (3) combining procedures that find their requirements in different clauses, and (4) combining documents across systems (i.e., ISO 14000 and ISO 9000).

6. Organizations may save time and effort by cross-referencing documentation, rather than including it in some documents. For example, an organization's EMS manual might refer to certain procedures that are properly documented elsewhere, instead of including them in the manual.

7. The structure of the ISO 14000 documentation system can be depicted as a pyramid divided into four horizontal components. These components, beginning at the top of the pyramid, are as follows: (1) Policy, (2) Procedures, (3) Practices, and (4) Proof. Although this is the usual structure, it is not the only possible structure. Any system that is compatible with ISO 14001, the organization's environmental aspects, and its everyday needs is an acceptable system.

8. ISO 14000 documentation may be maintained in electronic format. In fact, there are advantages to this approach (i.e., currency of documents, use of most up-to-date document). However, unless every person who needs to use the documentation has immediate access to the electronic format, hard copies should be made readily available.

## KEY CONCEPTS

| | |
|---|---|
| Clarity | Policies |
| Combining documentation | Practices |
| Cross-referencing documentation | Procedures |
| Documentation | Proof |
| Effectiveness | Reformatting documentation |
| Electronic documentation | Work instructions |

## REVIEW QUESTIONS

1. Define *document* and *documentation* as they relate to ISO 14000.
2. How does the concept of clarity relate to ISO 14000 documentation?
3. What is meant by effective documentation?
4. What elements should always be included in the written documentation for policies, procedures, and work instructions?
5. What is meant by reformatting documents? Why might an organization choose to do this?
6. Explain how combining documents might save an organization time when developing its documentation system.
7. Give an example that illustrates the concept of cross-referencing documentation.
8. Draw a simple illustration that depicts the typical structure of an ISO 14000 documentation system.
9. How does the concept of making documentation readily available to those who need it affect the use of electronic documentation?

## CRITICAL-THINKING PROBLEMS

The following activities may be assigned as individual, group, or discussion activities to be completed in class or out of class.

1. Develop an outline for the ISO 14000 documentation for a small manufacturing firm (518 employees).

2. Develop a cover page that is to be used with all ISO 14000 documents for the company in Problem 1. The cover page should ensure inclusion of the elements that should always be part of written documentation.

3. Locate a company in your community that documents its work procedures. Examine the documentation to identify instances in which procedures could be reformatted to become part of an ISO 14000 documentation package.

4. Using the documentation from Problem 3, identify instances where cross-referencing might decrease the amount of documentation needed.

## DISCUSSION CASE

The ABC Chemical Processing Company is preparing to seek registration as an ISO 14001 organization. Its top managers are discussing the best way to approach documentation. Two opposing points of view are noted. Three of the managers want to build a comprehensive stand-alone documentation system from scratch. The other three want to use every shortcut available, including reformatting, cross-referencing, and combining to minimize the amount of documentation. The company's CEO sees advantages and disadvantages to both approaches.

### Discussion Questions

1. Which approach do you think is better?

2. What would you advise the CEO about the company's documentation?

## ENDNOTES

1. Joseph Cascio, *The ISO 14000 Handbook* (Fairfax, VA: CEEM Information Services, 1996), p. 200.

2. ISO 14001, Annex A, Clause A.2.

3. Ibid., Annex A, Clause A.4.5.

# Registration and the Audit Process

## SELF-DECLARATION OR CERTIFICATION/REGISTRATION

ISO 14000 allows two levels of certification, as follows:

- **Self-Declaration**—A firm can implement ISO 14000 and *self-declare* that it is in conformance. **Self-declaration** is an internal scheme—involving no outside agency to confirm conformance. Rather, the organization uses internal audits of its EMS for that purpose. Variations on this scheme are possible. For example, the organization could hire an outside auditing firm (not a registrar), or a customer firm could audit the supplier firm for ISO 14000 conformance.
- **Certification/Registration**—In certification/registration the organization is formally certified by an accredited ISO 14000 registrar.

Once an organization determines that in its own best interest it should implement an environmental management system based on the ISO 14000 model, it must choose self-declaration or certification/registration. The organization may implement an ISO 14000-style EMS and self-declare conformance, or it may go further and have its EMS certified by an independent third-party registrar. The certification route will ultimately be the most used.

Certification is more common than self-declaration for several reasons. People worldwide agree that everyone needs to do the best job possible in protecting the environment at the global and local levels. If an organization declared that its environmental performance is excellent, conforming to ISO 14000, the assertion would have more credibility if an independent, knowledgeable authority confirmed it. No organization can compete without the support of customers; therefore, its claims need to be believable. In many cases organizations must have the support of the local community and political entities before they are permitted to operate or expand. In the United States and Canada, environmental concerns are being heard, and this situation can only become more sensitive over time as people become more environmentally informed.

Assume that you are a customer or an interested party. Would you give more credence to an organization that claims to be a good environmental citizen, or a competing organization whose environmental management system is accredited by a third-party registrar? You would need to consider which organization to buy from or support—the one you cannot be sure of or the one you know is environmentally responsible.

A self-declaring organization could be using poor judgment. The only rational reasons for an organization to stop short of registration, after going to the trouble and expense of implementing a conforming EMS, are cost and effort. Admittedly, registrars do not provide their services pro bono, but the organization would already have invested a lot of money and employee time in the development and implementation of the EMS. Further, it is true that the registrar's auditors will visit the organization every six to twelve months, and the visits may be seen as unwelcome and intrusive. On the other hand, to conform, even in a self-declared implementation, the audits must be performed by someone. Consequently, the intrusion and nuisance factors still will be present.

Finally, and probably most important, the registrar's personnel will be more objective in their audits than internal personnel. More critically, they will bring valuable training, experience, unbiased observation, and a powerful influence to the task, which may ultimately enable the organization to avoid serious and expensive environmental problems. Environmental problems can be far more expensive than the cost of a registrar.

---

### ISO 14000 INFO

## Certification or Registration?

*Throughout most of the world, the term* certification *is used in connection with environmental management systems. In this text* certification *and* registration *are used interchangeably. However,* registration *is the term preferred by registrars due to liability implications involving certifications in the United States.*

Whether the organization self-declares or uses a registrar, the same conformance requirements exist and environmental performance should be the same. The primary difference will be in what self-declaration versus certification means to the firm's stakeholders, principally its customers and employees. Certification by an accredited ISO 14000 registrar carries the full weight and prestige of the International Organization for Standardization. Self-declaration, on the other hand, is not likely to carry as much weight or prestige with anyone. This chapter assumes that the organization intends to seek formal registration through one of the accredited registrar firms.

# REGISTRATION PROCESS

Chapter 3 explained the requirements of ISO 14000, chapter 4 examined the elements of a sound EMS, and chapter 5 described the documentation and documentation system necessary to support the EMS. If the organization has followed through this far, it should be ready to start the **registration process**. There are eight sequential steps in the registration process, as follows:

1. Decision by the organization to conform to ISO 14000 and to seek registration.
2. Internal preparation by the organization to achieve conformance.
3. Internal determination that the organization has achieved conformance and that the EMS is functioning.
4. Accredited ISO 14000 registrar engaged to certify the organization.
5. Preliminary assessment and document review by the registrar.
6. Formal EMS audit and certification assessment by the registrar.
7. Elimination of nonconformances preventing registration.
8. Registration is awarded by the registrar.

Figure 6-1 illustrates the eight steps of the registration process in a flow diagram. The next several paragraphs provide a guided tour of Figure 6-1.

The first element of the diagram, Decision to Conform, is satisfied when the organization makes a commitment to conform to ISO 14000. This commitment includes the requirement to comply with all applicable environmental laws and regulations, a requirement of ISO 14000 (ISO 14001, Clause 4.2 c).

In the second element, according to ISO 14001, the organization must prepare its environmental policies, plans, procedures, and practices required for its EMS, and must communicate them to appropriate employees. It must identify environmental aspects and the real or potential impacts of the aspects. Also, it must document environmental objectives and targets that address the key aspects and impacts. The organization also must set up control processes for its environmental operations and documents. It must establish procedures and facilities for the maintenance of environmental records, for internal and external environmental communications, and for emergency preparedness and response. Finally, the organization must develop processes for checking and moni-

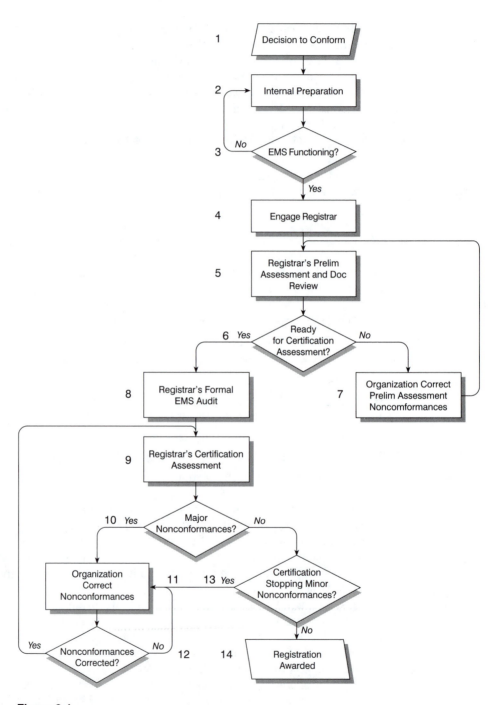

**Figure 6-1**
ISO 14000 Registration Process

toring its environmental operations, for making corrective actions when needed, and for reviewing the adequacy and effectiveness of its EMS.

It is a good idea for the organization to gain at least a preliminary feel for the suitability of its EMS and its constituent parts before it engages a registrar. The organization should define the employees' environmental roles, responsibilities, and authority, and conduct the training it finds necessary, and only then activate the EMS.

In the third element of the diagram, the organization asks, "Is the EMS functioning properly?" The organization should correct obvious malfunctions through its internal preparation loop (return to second element in diagram) before getting a registrar involved. Remember, the registrar cannot help an organization conform with ISO 14000; all that the registrar may do is determine whether or not the organization conforms. A conflict of interest would result if a registrar identified a particular problem and then proceeded to tell the organization how to correct it.

When the organization feels that the EMS is functioning properly, it is time to engage the services of a registrar, as depicted by the fourth element in the flow diagram. (Details on selecting a registrar will be discussed later in the chapter.)

As the fifth element element of the diagram illustrates, the registrar first will conduct a preliminary assessment (or initial assessment) and document review. Normally but not always this is accomplished in two stages. First, the registrar may request copies of critical EMS documentation to review before visiting the site. This action allows the registrar to alert the organization about any obvious areas that need attention prior to the visit, thereby helping the organization's preparation efforts. The actual preliminary assessment is nearly always done at the organization's site. Most registrars prefer to assess on-site because of the volume of material to be reviewed, the ease of asking questions and obtaining direct responses, and to get a sense of the organization's operation. Because different registrars use different methods, it is important for the organization to understand exactly how its registrar will conduct the preliminary and certification assessments before they occur.

During the preliminary assessment and document review, the registrar will, as a minimum, want to review the following:

- EMS manual, including environmental policy; policies addressing the ISO 14001 clauses; organization charts defining roles, responsibilities, and authority; emergency plans; improvement plans; and indices to other relevant documentation. (The registrar will want to review the constituent parts of an EMS manual if the organization has not developed a formal EMS manual.)
- The list of environmental aspects and impacts and the procedure for determining them
- Legal and regulatory requirements that apply to the organization
- Training programs, procedures, and records
- Internal audit reports
- Records of management review of the EMS

As the sixth element of Figure 6-1 illustrates, the registrar determines the organization's readiness for certification assessment. If the registrar finds nonconformances, which is usually true in preliminary assessments, the organization will be advised of areas that require additional work. The process then proceeds to the seventh element of the diagram, and a nonconformance correction-and-verification loop is developed using the seventh, fifth, and sixth elements. The organization is responsible for correcting the problems. Although the registrar will not (may not) provide substantive assistance, the organization is free to retain a consultant for such purposes, if it chooses to do so.

If the registrar finds no significant nonconformances (at the sixth element of the diagram) or if the organization has corrected to the registrar's satisfaction any problems that were noted in an earlier pass, the organization is considered ready for its formal EMS audit.

The eighth element of the diagram reflects that the registrar will conduct a formal audit of the organization's EMS. (The audit process is described later in this chapter.)

Next, at the ninth element, the audit report will be used by the registrar to assess whether certification should be granted.

As the tenth element of the diagram indicates, if there are any major nonconformances, registration cannot be granted until they are eliminated. It would be unlikely, though not impossible, that a major nonconformance would exist at this point because of the preliminary assessment previously completed by the registrar. It would be most embarrassing for the registrar to uncover a major nonconformance at this point. Nonetheless, if any major nonconformances are discovered, the organization implements corrective action (eleventh element of diagram) and verifies the corrective action (twelfth element). When the organization is satisfied that the nonconformance is corrected, it reverts back to the registrar's certification assessment (ninth element), providing the registrar with evidence of conformance. At this point the registrar usually revisits the site to confirm the elimination of major nonconformances (tenth element).

Once any major nonconformances are eliminated, the certification process advances to the thirteenth element of the diagram. Minor nonconformances do not necessarily cause the registrar to withhold registration. However, registration may be withheld if (1) there are minor nonconformances that fall into a pattern which suggests that an element of the EMS is not effectively implemented, or (2) too many minor nonconformances exist, suggesting that the organization has not taken the EMS design or implementation seriously. If registration is withheld for either of these cases, the minor nonconformances must be corrected and verified by the organization (eleventh and

---

**ISO 14000 INFO**

## Major and Minor Nonconformances

*A **major nonconformance** results from (1) a failure to fulfill any requirement of ISO 14000 or (2) multiple minor nonconformances sufficient to lead auditors to conclude that an ISO 14000 requirement is not effectively implemented.*

*A **minor nonconformance** is a single lapse in implementing an EMS requirement. It is seen as an anomaly, rather than a systemic issue.*

twelfth elements in diagram) and then verified by the registrar through the final verification-and-certification assessment loop (ninth, tenth, and thirteenth elements) *before registration can be issued*. Depending on the nature of the nonconformances, the registrar may schedule another site visit in order to verify correction.

Once no major nonconformances or certification-stopping minor nonconformances are present (tenth and thirteenth elements), the registrar may award registration, as reflected in the fourteenth element.

The ISO 14000 registration process can lead to three possible outcomes:

- **Approval**—The EMS is found to be in substantial conformance with ISO 14001 requirements. In this case a document certifying registration to ISO 14000 is issued to the organization by the registrar.

- **Conditional approval**—All elements of ISO 14001 have been addressed but either they are not fully documented, not fully implemented, or objective evidence leads the auditors to believe there is a systemic problem. These issues must be corrected by the organization and reviewed by the registrar before full approval can be granted. The registration certificate will be withheld until the issues are satisfactorily closed.

- **Disapproval**—At least one requirement of ISO 14001 has not been addressed by the organization, or the EMS is demonstrated to be ineffective at meeting policy commitments and/or objectives and targets.

## SELECTING A REGISTRAR

While an organization prepares for ISO 14000 registration, one of its most important steps will be selecting a registrar. This step is important for several reasons, including cost (which can vary widely), recognition (by interested parties), familiarity with the organization's industry (it helps if the organization and the registrar speak the same technical language), and integrity. The best policy is to consult with registered firms to solicit information about their registrars. Registrars should provide references that can be checked. The organization is solely responsible for choosing a registrar. It will spend a significant amount of money for the registrar's services, to say nothing of the internal cost of developing and implementing the EMS and its supporting infrastructure. Consequently, the investment should be protected by securing a registrar who is competent, whose services are fairly priced, and whose integrity is beyond question.

The issue of integrity is a double-edged sword. Perhaps a registrar has the reputation for granting certification easily, which, on the surface, might seem like an advantage. However, two problems exist with this approach. First, even if the organization can secure an "easy" registration with a sub-par EMS, although the ISO 14000 certificate may be on the wall, the organization will still be struggling to achieve the environmental performance required by law and regulation, desired by the organization itself, and increasingly demanded by the community and customers. The organization should develop and implement the best EMS it is capable of in order to more easily and consis-

## ISO 14000 INFO

*ISO 14000 registrar firms conduct certification audits and issue certificates of registration to organizations having environmental management systems that conform to ISO 14001. The registrar is an independent auditing firm that is competent to perform EMS audits and determine EMS conformance and whose systems conform with international guidelines (including ISO 14010, 14011, and 14012). To perform EMS audits, evaluate conformance with ISO 14000, and issue certificates of registration, the registrar itself must be accredited by one of 40 accreditation bodies worldwide. In the United States ISO 14000 registrars are accredited jointly by ANSI-RAB (American National Standards Institute and the Registrars Accreditation Board). Thirty-nine other countries have their own registrar accreditation bodies, including:*

- *Canada, Standards Council of Canada (SCC)*
- *France, Comité Francaise d'Accreditation (COFRAC)*
- *Germany, Deutscher Akkreditierungerat*
- *Ireland, Irish National Accreditation Board*
- *Japan, Japan Accreditation Board*
- *Mexico, Dirección General de Normas*
- *The Netherlands, Raad voor Accreditatie (RvA)*
- *United Kingdom   United Kingdom Accreditation Service (UKAS)*

tently achieve the required environmental performance. This eliminates the need to seek a registrar with a reputation for granting certification easily.

Second, if one organization can identify a registrar with less stringent certification standards, so can other organizations. In fact, such a registrar's reputation is probably already known throughout the world. Registration from a disreputable registrar is like having a degree from a disreputable college. Your registration will have no credibility among customers, competitors, legal and regulatory agencies, and other interested parties.

Firms should choose registrars based on these criteria:

- Past ethical performance
- Familiarity with the industry
- Reputation for competence, fairness, and objectivity
- Cost

Another factor to consider when selecting a registrar is availability of the registrar to suit the organization's schedule. In addition, the organization may also consider professional chemistry, or the likelihood of being able to develop a good working relationship.

# EMS AUDIT—DEFINITION

ISO defines *environmental audit* as follows:

> Environmental audit—systematic, documented verification process of objectively obtaining and evaluating audit evidence to determine whether specified environmental activities, events, conditions, management systems, or information about these matters conform with audit criteria, and communicating the results of this process to the client.[1]

From this definition we can determine that an **environmental audit** has the following characteristics:

- They are systematic—not improvised, not casual.
- The audit process is documented and will be performed accordingly.
- The audit team will objectively seek and evaluate information and evidence concerned with environmental activities, events, conditions, management systems, and related information.
- The objective is to determine whether these activities, events, conditions, management systems, and related information conform with the audit criteria.
- Results of the audit process are communicated to the client (organization commissioning the audit).

These characteristics are logical and straightforward. However, of the fourth point one might query, What are the audit criteria? Who establishes them? Again, according to ISO, *EMS audit criteria* are:

> ... policies, practices, procedures or requirements, such as those covered by ISO 14001 and, if applicable, any additional EMS requirements against which the auditor compares collected evidence about the organization's environmental management system.[2]

In effect, the EMS audit criteria are the organization's own statements of intent. The policies, practices, and procedures are those developed by the organization. Requirements may originate from standards (such as ISO 14001), guidelines (such as ISO 14004), the organization itself, and legislative or regulatory agencies. The registrar will plan the audit around the organization's EMS documentation and will confirm the audit criteria with the organization prior to an audit.

In its EMS—with policies, procedures, practices, records, and legal and regulatory requirements—the organization has told the registrar how it intends to operate and maintain its environmental activities. Through the EMS audit, the registrar will verify whether the organization is doing what it said it would do (conformity) and whether the results confirm that the EMS is effective.

# AUDITORS

The following paragraphs will explain several topics dealing with the different kinds of environmental auditors, their qualifications, audit teams, and auditor certification. We begin by examining the need for environmental auditors.

## Why Auditors are Needed

ISO 14000 is a series of international standards for environmental management systems for organizations whose operations have environmental aspects. In a perfect world it might be sufficient to publish an environmental standard and expect organizations around the world to conform on their own for the benefit of their customers, themselves, and the community; certification would be unimportant. Unfortunately, our world is far from being perfect. Even in an imperfect world, organizations would not need a certification process if there were a single world authority that could mandate adoption of the standard and enforce adherence. The International Organization for Standardization (ISO) has no authority to mandate or enforce its standards; its job is to develop standards through its international committees. Individual organizations decide whether or not to use the standards. Any organization worldwide is free to adopt or not adopt ISO 14000. The decision is inherently an internal one, although organizations may feel pressure from customers, government agencies, and other interested parties.

## Internal and External Auditors

Organizations seeking to use the ISO 14000 EMS model need "independent" observers to verify conformance with the EMS and its constituent elements. These independent observers are the auditors. Auditors include employees of the organization and employees of a third-party registrar firm. While it is not normally a problem for third-party auditors to be independent, objective observers, it is an issue for internal auditors. Internal auditors should never audit their own organizational component (i.e., department, function, group). For example, an engineer should never audit the engineering department to which he or she is assigned. An auditing watchword states that internal auditors must be independent of the organizational elements they audit.

 ## Auditor Qualification

ISO 14012 details the **qualifications** for environmental auditors. Qualification criteria include the following:

- Education—at least secondary (high school) or equivalent
- Five years' work experience (four years if holding a college degree) contributing to skills and understanding in some or all of the following:
  1. Environmental science and technology
  2. Technical and environmental aspects of facility operations

3 ● • Relevant requirements of environmental laws, regulations, and related documents
4 ● • Environmental management systems and standards against which audits may be conducted
5 • Audit procedures, processes, and techniques (See ISO 14012, Clause 4, for exceptions and variations.)

■ Formal training on the subjects in the five points above
■ On-the-job training involving 20 audit work days
■ Personal attributes and skills should include the following:[3]

  • Competence in clearly expressing concepts and ideas, orally and in writing
  • Diplomacy, tact, ability to listen
  • Ability to maintain independence and objectivity
  • Personal organization skills
  • Ability to reach sound judgments based on objective evidence
  • Sensitivity to the conventions and culture of the country or region in which the audit is performed and of the organization being audited.

■ Auditors must maintain relationship of confidentiality and discretion with client[4]
■ Lead auditors have same qualification criteria but must have additional experience and demonstrated leadership capability (See ISO 14012, Clause 8, for further information.)

The auditor qualification criteria just explained were developed by ISO for auditors who work for the registrars. ISO contends that "internal auditors need the same set of competencies, but may not meet in all respects the detailed criteria [listed above] depending upon such factors as the size, nature, complexity and environmental impacts of the organization; and the rate of development of the relevant expertise and experience within the organization."[5] This means that the organization's employees may be used as internal auditors without meeting all of the qualification criteria. However, no employee should be allowed to audit an EMS without first receiving training on the fundamentals of environmental auditing. In addition, as stated previously, internal auditors must be independent of the function being audited.


## Environmental Audit Teams

The registrar's **environmental audit team** will typically be comprised of several certified environmental auditors and a lead auditor. The number will vary according to the size of the organization being audited, and can in rare cases be a single individual. The lead auditor considers the following when selecting members for the audit team:[6]

■ Qualifications of potential auditors
■ The type of organization, processes, activities, or functions being audited
■ The number, language skills, and expertise of the audit team members

- Any potential conflict of interest between the audit team members and the organization being audited
- Requirements of clients and certification and accreditation bodies

Most registrars attempt to include as team members subject matter experts. These are people with specific kinds of experience beyond ISO 14000. For example, if a facility that manufactures electronic equipment is to be audited, the audit team might include one or more people who are experts in such an environment. Some registrars, however, believe that a certified environmental auditor, regardless of specific technical experience, can satisfactorily audit the EMS of any kind of organization.

Occasionally audit teams may include technical experts who do not serve as auditors. These experts assist the auditors in areas (processes) outside the realm of the auditors' technical expertise. This is a compromise or middle-ground position when the registrar's auditors are not familiar with the specific processes of the organization being audited.

The audit team may occasionally have one or more auditors-in-training. These are auditors who are accumulating environmental audit hours in order to obtain full certification.

## Auditor Certification

Auditors working for registrar firms must be certified as meeting the ISO environmental auditor qualification criteria. Environmental auditors who work for registrars must demonstrate appropriate experience and qualifications as described earlier and must complete a course and pass a comprehensive examination for ISO 14000 auditors. In the United States, environmental auditor certification is provided by the Registrars Accreditation Board (RAB). Other nations use similar auditor certification methods.

# TYPES OF EMS AUDITS

Three general types of EMS audits are:

- Registration audit
- Surveillance audit
- Internal audit

## Registration Audit

The **registration audit** is performed by the registrar firm that is hired by the organization seeking registration. The intent of this audit is to verify to the registrar initial conformance with ISO 14000 and the organization's EMS. Registration audits are repeated every three years. If a registration audit is completed satisfactorily, the organization becomes registered.

The initial registration audit begins with two time-phased elements. The first element is a documentation review by the lead auditor (usually) at the registrar's place of business.

This involves a detailed review of the organization's EMS documentation; the documentation is compared against the requirements of ISO 14001 and other relevant requirements.

The second element is an optional on-site preliminary assessment of the organization's readiness for registration. The preliminary assessment should be viewed as a trial run for the registration audit. The trial run allows the auditors to gain first-hand knowledge of the organization's operations and its personnel. From the preliminary assessment, significant gaps in the EMS implementation can be identified so that the organization can close the gaps before the registration audit.

The final phase, the actual registration audit, is performed after the registrar determines that the organization is prepared. The registration audit is conducted at the organization's site. Areas audited include:

- **Conformance audit**—audit policies, procedures, practices, proof; against ISO 14001, legal, and regulatory requirements
- **Conformance audit**—audit practices, proof; against the elements of the EMS
- **Process audits**—audit conformance with EMS practices, work instructions
- **Compliance audit**—audit policies, procedures, practices, proof; against legal and regulatory and other requirements

Figure 6-2 is a representation of an EMS audit.

Should the registrar find nonconformances that cannot be corrected while the audit team is on-site, one or more follow-up audits may be required to verify subsequent elimination of the nonconformances.

## Surveillance Audit

A **surveillance audit** is also performed by the registrar to verify continued conformance with ISO 14000 and the organization's EMS. These audits, depending on the individual registrars, are conducted at six- or twelve-month intervals. Satisfactory surveillance findings result in continued registration of the organization.

---

### ISO 14000 INFO

*The following are items to remember for compliance audits:*

1. *Compliance to legal, regulatory, and other relevant requirements is required by ISO 14000.*

2. *The registrar will audit for compliance to only those laws and regulations which have been* identified as applicable by the organization.

3. *Although compliance to applicable laws and regulations is an ISO 14000 conformance requirement, compliance will not be the main focus of the registration or surveillance audits.*

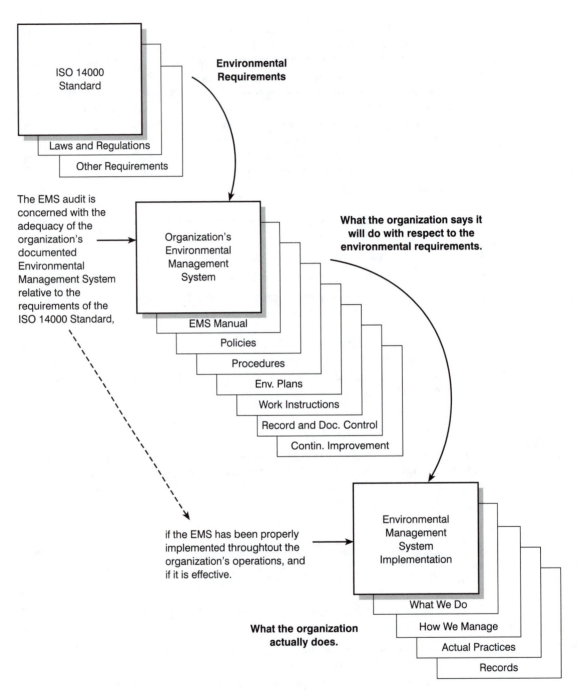

**Figure 6-2**
EMS Audits—By Registrar or Internal Auditors

## Internal Audit

An **internal audit** is conducted by audit teams comprised of employees from the organization. These audits are typically performed between visits by the registrar's auditors in order to verify to management that nonconformances have been corrected and that the EMS is working effectively. However, the organization may increase the audit frequency in areas where nonconformances have been noted or where extra attention is needed. Internal audits focus on the same areas as the registrar's audits, but the audience is the organization's management team. Satisfactory results in an internal audit should mean that no nonconformances will be noted at the next surveillance audit.

Internal audits are also performed prior to the initial registration audit to confirm an organization's readiness for registration.

# OBJECTIVES OF EMS AUDITS

If you ask someone why he or she goes to college, that person is likely to respond "I'm going to college to earn a degree in ———— " However, just earning a degree is not the real objective of four or more years of study. The real objective of going to college may be to enter (and be successful in) a prized career, be better able to provide for one's family, have a better chance of achieving a secure future, or develop personally. The work behind the degree prepares one to pursue a real goal and is therefore only a means to the desired end.

EMS audits are similar. On the surface, audit objectives would seem to be obvious. For example, most organizations would declare that their objective for a certification audit is registration. Similarly, the objective of surveillance audits may be seen as retaining registration. Both are reasonable, valid objectives. Certainly, though, that is not the end. Surely there is more than having the newly earned or updated ISO 14000 registration certificate displayed in the organization's lobby. Like an individual's degree displayed on an office wall, the registration certificate is merely a symbol signifying that development and improvement have occurred. The real importance is that the improvement has allowed the organization to achieve certain objectives, such as:

- Confirmation by an independent, knowledgeable, objective party that the organization's EMS is effectively implemented and can be expected to aid the organization in meeting its legal and regulatory requirements and its own environmental performance objectives.

- Improvement (or confirmation) of its image and, hence, its business by being able to legitimately claim to customers, employees, legal and regulatory agencies, and other interested parties that it is an environmentally responsible organization.

- Knowledge that the organization is on a sound environmental path that should continue to improve performance and avoid disasters.

A registrar and its audit team have several objectives in performing audits. They include:

■ Perform audits objectively, without bias.

■ Determine conformance with ISO 14000 and the organization's EMS.

■ Determine that the EMS has been properly implemented and operated (maintained).

■ Determine management's ability and willingness to review EMS processes for continued suitability and effectiveness and for continuous improvement.

■ Identify improvement opportunities.

Registrars should have no objective relating to the organization's certification. Rather, they should ensure that audits are conducted with objectivity, fairness, confidentiality, and integrity. Auditors, in turn, may be indifferent concerning an organization's objective for certification. This may sound harsh, especially since the organization must pay a significant fee to the registrar for the audit. However, if the integrity of ISO 14000 certification is to be maintained, auditors must remain independent and objective, guarding against any possibility of disregarding or compromising standards.

## AUDIT SCOPE

According to ISO 14011, Clause 5.1.1, the **audit scope** establishes the boundaries of the audit. It specifically establishes:

■ Physical locations to be visited

■ Organizational activities to be audited

■ Resources required for the audit

  • Audit team

  • Organization's audit support requirements

  • Physical resources; i.e., audit team work space

■ Manner of reporting on the audit

The formal audit scope is developed by the lead auditor in consultation with the client, normally the organization to be audited. Consequently, the organization knows beforehand what the auditors will be looking at, how many auditors will participate, how long the auditors intend to be on-site, and what resources the organization must provide.

The scope of the initial registration audit, as well as the reregistration audit every third year, must cover the organization's entire EMS. The audits will include an examination of the environmental policy, EMS manual, policies, plans, procedures, work instructions, and documented requirements for records. Collectively these documents tell the auditor what the organization plans to do and how it intends to do it. The auditor then determines by interview, observation, and through the examination of records the degree of effectiveness of the EMS implementation. In other words, the auditor determines to what degree the organization is doing what it said it would do and whether it is doing it how it said it would do it. The auditor will be interested in any activity that has environmental aspects, and certainly any activity specified in ISO 14001 or in the EMS documentation will be within the scope of the audit.

The annual or semiannual surveillance audits are not as far-ranging but may probe any environmentally related area of interest or concern to the auditor. These audits usually focus on randomly selected target areas together with any nonconforming activities that were noted during the previous audit or that may have been brought to the attention of the registrar.

In summary, audits may involve any part of the organization that is specified in ISO 14001 or in the organization's EMS documentation or any activity that may in any way affect the environment. Registration and reregistration audits are comprehensive. Surveillance audits may be as comprehensive as the registrar thinks necessary and appropriate.

## AUDIT PROCESS

Both internal and third-party audits by the organization's registrar are based on the requirements of the ISO 14001 standard and are governed by the principles and procedures of ISO 14010 and 14011, respectively. The auditor's checklist will contain items from every ISO 14001 Section 4 clause. The **audit process** will attempt to determine the following two things:

- The adequacy of the EMS documentation prepared in response to the requirements set forth in the Section 4 clauses of ISO 14001.
- The effectiveness of the EMS implementation, that is, the degree to which the relevant activities conform to the elements of the EMS, including policies, procedures, practices, work instructions, hierarchy, staffing, records, and legal, regulatory, and other requirements.

### Audits Performed by Registrars

Registrars perform registration audits and surveillance audits. In the section that follows we have concentrated on the registration audit because it is the most pervasive. However, we have included notes to indicate applicability or differences relative to the two audit types.

Preaudit Activity

*Note*: Applies only to the initial registration audit.

The registrar requires copies of critical EMS documentation in advance of site visits. This helps in planning the registration audit and is less expensive than an on-site review of the materials. In reviewing these copies, the registrar can note areas in the EMS documentation that need attention before the on-site visit.

Preliminary Assessment Visit (Optional)

After having completed a study of the organization's EMS documentation, the registrar may, before the initial registration audit, send a representative, normally the leader of the audit team, to the organization's facility in order to develop a sense of the operation. This step, referred to as the **preliminary assessment visit**, is useful for the final planning of the audit—plans which specify the composition of the audit team, number of team members, and number of days required for the audit. The plans are also helpful for identifying obvious areas of concern that will be shared immediately.

### Preparation for the Audit

*Note*: Applies to registration audits, including reregistration.

The audit will be scheduled for a date on which the organization and the lead auditor agree. Before the date, the organization should have completed all preparations for ISO 14000, and all employees should have fully prepared for the audit. Employee preparation should include details of what to expect as well as proper conduct when dealing with the auditors. Employees should be cooperative, open, nondefensive, and nonargumentative. In addition, they should answer all questions truthfully. On the other hand, employees should not attempt to answer questions if they do not know the answer. In these instances, a simple "I don't know" is the best response. Employees should not disclose information that is not requested. It does neither the organization nor the auditors any good when employees at any level volunteer information. The auditors look for specific facts, and they know the questions to ask in order to obtain these facts. (*Note*: Employee conduct should be the same for all types of audits.)

### Opening Meeting

*Note*: Applies to registration and surveillance audits, although surveillance audits will generally be shorter and require less support.

Immediately before the audit begins an **opening meeting** will be held, involving the audit team and key personnel from the organization being audited. During this meeting, which usually lasts one hour, the lead auditor will accomplish the following tasks:

- Emphasize the positive nature of the audit (i.e., the team is present by invitation to enable the organization to be registered to ISO 14000, not to prevent that outcome).
- Introduce members of the audit team to the organization's key staff.
- Explain the scope, objectives, plan, and timetable of the audit, making certain that everyone present understands.
- Emphasize the importance of timely employee participation.
- Explain the mutually agreed audit criteria (ISO 14001, the EMS, and supporting documentation).
- Summarize audit procedures and methods.
- Review confidentiality issues (i.e., nonattribution for information gained during interviews, registrar's confidential treatment of all information).
- Discuss the types of findings, major and minor observations, and their impacts on the audit outcome.
- Acquaint the organization with the auditing assignments (i.e., who will review specific items).
- Review any relevant safety and emergency procedures that may be appropriate.

- Confirm availability and location of required resources and facilities (employees to be interviewed, guides, phones, reproduction machines, team workspace).
- Establish time for end-of-day reviews and for the audit closing meeting.
- Respond to any questions.

### Facility Tour

*Note*: May not be required for surveillance audits, depending on the auditors.

It is customary for the organization to conduct a tour of the facility for the auditors. This **facility tour** allows the audit team to familiarize itself with the location of the various activities and the locations of employees to be interviewed. It also enables the auditors to develop questions they had not considered before seeing the activities in operation. Depending on the size of the organization, this tour could require as little as one hour or as long as half a day. If the tour could not be completed in half a day, the audit team should be divided to tour only their specific areas of interest for the audit.

### Registration Audit

*Note*: The surveillance audit generally seeks to confirm continued conformance and continued effectiveness of the EMS, including any changes introduced through corrective or preventive action. Therefore, it is a more narrowly focused audit.

The purpose of the registration audit is to determine objectively if the organization conforms to the requirements and intent of ISO 14001, its own EMS, and other EMS-related documents. Determining conformance requires the following:

- The EMS document system must be complete, responsive to ISO 14001, and comprehensive.
- The total EMS must have been implemented.
- The EMS implementation must be shown to be effective.

The first requirement in this list may be satisfied by comparing the EMS documents with the requirements of ISO 14001. The second and third requirements must be satisfied through objective evidence, which should be collected by the audit team.

Depending on the size and nature of the organization, the audit may last up to one week. The audit team will spend approximately 25 percent of its time examining EMS documentation. The remainder of the time will be spent collecting evidence to verify operational conformance with the EMS. Remember, the EMS tells the auditors (as well as the employees) what the organization says it is going to do. The auditors, then, must verify that the organization is actually doing what it said it would do. A major nonconformance will stop the registration, and perhaps the audit itself. Minor nonconformances, unless the numbers reach the lead auditor's threshold or unless they are grouped in such a way as to indicate that the EMS is not effectively implemented, will neither stop the audit nor cause the registration to be withheld. Often minor discrepancies can be remedied instantly (although corrections will require documentation).

The auditors try to keep the organization informed of their findings through the course of the audit, but inevitably some audit observations or minor nonconformances can be revealed only at the exit meeting after the auditors have concluded their analysis of all the data. In order for the process to continue, the organization and the auditors must agree to a formal schedule for taking **corrective action** on any open issues. Figure 6-3 is a flow diagram of the corrective action process.

### End-of-Day Review

*Note*: The end-of-day practice is usually observed for both registration and surveillance audits.

At the end of each audit day, the team holds a short meeting (thirty to sixty minutes), referred to as **end-of-day review**, with the organization's key personnel to discuss the information learned during the day. The three purposes of the end-of-day meetings are:

1. Allow the organization to correct nonconformances during the audit.
2. Keep the organization informed concerning findings and progress of the audit and to mitigate any administrative difficulties.
3. Minimize, or prevent, surprises at the exit meeting.

Nonconformances are discussed at the end-of-day meeting. The organization may correct the nonconformances during the audit or, if this is not possible, propose a schedule for correcting them.

### Exit Meeting Preparation

*Note*: The exit meeting applies to registration and surveillance audits. Techniques used are the same.

Once its investigative tasks have been completed, the audit team assembles to prepare for the exit meeting. Collected evidence is discussed, evaluated, and grouped into three categories:

- Conforming—no action required by the organization
- Nonconforming—the organization must address
- Observations and concerns—the organization may or may not take action

The auditors will confirm all input with evidence and reach a consensus regarding all nonconformances and recommendations. If insufficient evidence is available on a particular point, the audit team may present an observation or concern.

Audit findings, showing areas of conformance or nonconformance, will be summarized, documented, and prepared for presentation to the organization's key staff at the exit meeting. Each nonconformance identified is numbered and documented on nonconformity report (NCR) forms along with a statement of the requirement not satisfied. A typical NCR is shown in Figure 6-4. Major nonconformances prevent registration; minor nonconformances usually do not.

**Figure 6-3**
Corrective Action Process

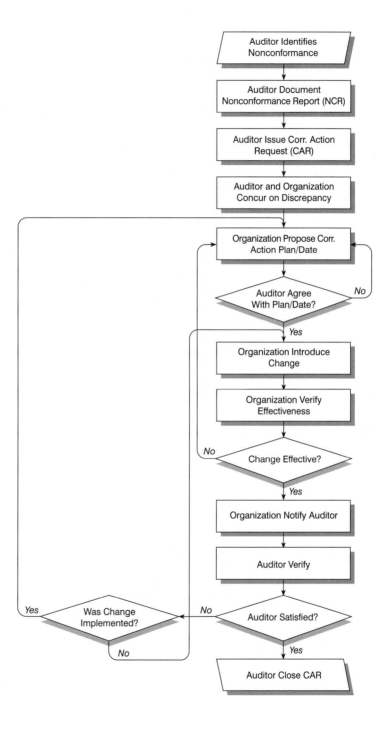

**EMS Nonconformance Report**

**SDC**

AUDIT FOR (FIRM):

NCR #

ACTIVITY:

DATE

ISO 14001 CLAUSE #

REF. DOCUMENT:

NONCONFORMANCE:

☐ MAJOR  ☐ MINOR  ☐ OBSERVATION

RESPONSIBLE MANAGER    DATE

AUDITOR    DATE

**Figure 6-4**
Typical EMS Nonconformance Report Form

The nonconformances also are documented on corrective action request (CAR) forms. A typical CAR form is shown in Figure 6-5. CARs are signed by both the auditor and the organization's manager in charge of the relevant activity. In addition, the responsible manager must sign off on the date that has been agreed to for completion of corrective action.

### Exit Meeting

*Note*: The surveillance audit exit meeting is generally the same as for the registration audit, but shorter.

The purpose of the **exit meeting**, which can last up to four hours, is for the audit team to present its findings. Employees who should attend the exit meeting include key management staff and managers or supervisors of the functions audited. A typical agenda for the exit meeting is as follows:

- Reintroduce the audit team
- Review scope and purpose of the audit
- Review briefly how the audit was conducted
- Review the audit criteria; i.e., ISO 14001, the organization's EMS documentation, and other related documentation
- Review nonconformance levels
  - Major—failure to meet a requirement of ISO 14001
  - Minor—a single lapse
- Summarize findings of conformance, presented by lead auditor
- Present nonconformances (including those which were cleared during the audit), presented by auditors
  - Obtain assurance that findings are understood by the organization
  - Obtain acknowledgment from the organization that findings were based on objective evidence
  - Resolve any disagreements
  - Agree on corrective action schedule
- Disclose the recommendation of the team (for the registrar) in regard to registration, presented by lead auditor

Copies of the NCRs and CARs are released to the organization, usually with a preliminary copy of the final report containing factual statements of discrepancies and objective evidence in support of findings. (*Note*: This assumes that the organization is the registrar's client.)

## Internal Audits

Internal EMS audits are a requirement of ISO 14001, Clause 4.5.4, in support of Clause 4.6, which is concerned with management review of the EMS to assure its continuing suitability and effectiveness. The audits must be periodic, performed according to a

**EMS Corrective Action Request**

SDC

AUDIT FOR (FIRM) | CAR #

ACTIVITY | DATE

ISO 14001 CLAUSE # | REF. DOCUMENT

AUDITOR | FIRM REP.

DATE | DATE

DESCRIPTION OF PROBLEM: (ADD SHEETS AS REQ'D.)

CORRECTIVE ACTION PLANNED:

PLANNED COMPL. DATE:

FIRM REP SIGNATURE | DATE

AUDITOR'S FOLLOW-UP DETAILS:

SATISFACTORY?  [ ] YES  [ ] NO  (CHECK ONE)

LEAD AUDITOR SIGNATURE | CAR CLOSE OUT DATE

**Figure 6-5**
Typical EMS Corrective Action Request Form

schedule, rather than on an ad hoc basis. The purpose of the internal audits is exactly the same as audits performed by a registrar; that is, to determine (1) whether the EMS conforms to the requirements of ISO 14001 and other planned arrangements and (2) if it has been properly implemented and maintained. Results of the internal audits are released to management.

Schedules for internal audits should consider the importance of the activities concerned, from the point of view of environmental aspects, and results of previous internal and external audits.

The process used is similar to that described earlier for registrar audits, but may not require the formal opening meeting or the end-of-day review. An exit meeting should be held with management and the supervisors of the activities audited.

Nonconformances must be documented in a manner similar to the registrar's process. Schedules for corrective action must also be documented and signed off by the manager of the relevant activity.

Obviously, internal audits cannot result in certification or withholding of registration. However, in the case of self-declaration of conformance with ISO 14000 (i.e., when used in lieu of actual registration), the internal audit is the basis for declaring conformance or nonconformance.

# FOLLOW-UP ACTIVITIES

## Follow-Up Activities—Registrar Audits

*Note*: Follow-up activities will be essentially the same for registration and surveillance audits.

### Audit Reports

The formal **audit report** is written by the lead auditor after the audit has been completed. This report will contain the elements presented, and provided to the organization in draft form, at the exit meeting. There should be no surprises; anything included in the report should have been presented during the exit meeting.

The audit report is compatible with the agreed audit plan. If during the course of the audit an auditor went beyond the scope of the audit plan, any resulting findings should not appear in the report. On the other hand, if a finding resulting from efforts beyond the scope of the audit plan is relevant, it should not be ignored. In such cases it is best to inform the organization's management through an informal channel. At the very least, the lead auditor should be certain to include the issue in the scope of the next audit.

Through coordination between the lead auditor and the client, distribution of the audit report is established in the audit plan. In most cases the client and the organization are one and the same. Occasionally the client will be a second-party firm, possibly an attorney. The audit report normally is distributed to the client, whether the client is the organization, its legal representative, or some other entity that contracted for the audit. To illustrate the latter, a corporate headquarters in Cleveland could contract a registrar to audit one of its divisions in Baltimore. The lead auditor will distribute the report in accordance with the client's instructions, regardless of the client's identity.

Confidentiality regarding the audit report must be maintained on two levels. First, the lead auditor must never distribute the report to any third party, unless directed to do so by the client. Second, statements used as evidence in the audit process must not be attributed to individuals in the organization.

### Corrective Action

Open nonconformance reports and corrective action requests must always be completed; otherwise, there is nothing gained by auditing nonconformances. Refer to Figure 6-3 to review the corrective action process. Note that no NCR/CAR resulting from a registrar's audit is considered closed until the lead auditor signs off indicating a satisfactory closeout. In some instances it will be necessary for the auditor to make a special visit to the site to verify satisfactory closure.

The lead auditor tracks the outstanding CARs against their scheduled completion dates. When dates are missed, the lead auditor *may* initiate action to determine the reason. Another lead auditor (or registrar) may not do so, however, believing that an organization wanting registration or wanting to retain it will take care of corrective actions without prodding from the registrar. This is not an unreasonable attitude; the registrar is not in the business of managing the activities of the organization. For this reason the organization should always be proactive in taking the initiative to keep the lead auditor informed of progress and problems associated with completion of CARs. Every reasonable attempt should be made to meet the completion dates agreed to, but when it becomes apparent that a date will be missed, it is prudent to inform the lead auditor or the registrar immediately with an explanation. As long as the organization is working in good faith to clear CARs and there is no pattern of inattention or disinterest, negative consequences are not likely.

### Awarding or Withholding Registration

Following the registration audit, the registrar's review of the lead auditor's report, and the recommendation of the audit team, the registrar may do the following:

1. **Award registration**—if no major nonconformances were found, if there was no pattern of minor nonconformances to indicate failure in the EMS implementation, and if the number of minor nonconformances was acceptably low.

2. **Award conditional registration**—pursuant to timely and effective resolution of specified minor nonconformances.

3. **Withhold registration**—until the organization demonstrates to the lead auditor's satisfaction that major nonconformances and/or patterns of minor nonconformances and/or an excessive number of minor nonconformances no longer exist.

*Note*: For surveillance audits, the team's recommendation may be to continue or discontinue registration.

### Preparation for the Next Audit

*Note*: This applies to both registration and surveillance audits.

The lead auditor will ensure that any areas of concern from the just-completed audit are targeted for follow-up at the next audit. Areas of concern include the following:

- Any activity that resulted in a reported unfavorable observation.
- Any area that had nonconformance reports and corrective action requests, even though implemented.
- Any area that resulted in an informal negative report (as in the earlier example).
- The effectiveness of all completed corrective actions.

The organization should be aware that the next audit will inevitably emphasize areas that demonstrated weaknesses in the EMS on the last audit, so that the organization can prepare appropriately.

## Follow-Up Activities—Internal Audits

Follow-up activities from internal audits generally follow the same scheme outlined for registrar audits, with the exception that internal audits are not concerned with awarding registration.

### Audit Reports

At the conclusion of an internal audit, information on the results of the audit must be provided to the organization's management.[7] Information from internal audits forms one element of management's review of the EMS's continuing suitability, adequacy, and effectiveness.[8]

### Corrective Action

The internal corrective action process is the same as that involving the registrar's auditors. In Figure 6-3 the word *activity* can be substituted wherever *organization* is used. This creates a flow diagram that can be used to deal with internal NCRs and CARs. *Auditor* in the flow diagram becomes *internal auditor*.

Nonconformances discovered during internal audits must be documented in the same manner as those of registrar audits, which is through a system of NCRs and CARs. The major difference is that completion of corrective action will be signed off by the internal lead auditor rather than the registrar's lead auditor. The forms used should be similar to those shown in Figures 6-4 and 6-5 but tailored for internal use.

Corrective action requirements stemming from internal audits should carry the same weight as those from an audit performed by the registrar. Internally generated NCRs and CARs become part of the permanent environmental record system,[9] and are reviewed by the registrar's auditors. If corrective action is incomplete or ineffective, the organization will be faced with a nonconformance by the registrar.

The internal lead auditor should track completion of corrective actions against the agreed schedule. The internal lead auditor should be actively engaged in pushing for on-time completion and advising both the activity and the organization's management of insufficient attention to corrective action requirements as well as schedule slippage.

### Preparation for the Next Audit

Like registrar audits the next internal audit should be influenced by past performance. If an activity has demonstrated weakness in the EMS or its implementation, that activity should be targeted in the next audit. In discussing internal audits, ISO 14001, Clause 4.5.4, says in part,

> The organization's audit program, including any schedule, shall be based on the environmental importance of the activity concerned and the results of previous audits.

Planning for internal audits should always include information from registrar audits. For example, corrective action resulting from a registrar's audit should be internally audited for proper implementation and effectiveness before the next registrar audit. By doing so any lingering difficulties may be eliminated before the registrar's next visit.

## SUMMARY

1. An organization may implement ISO 14000, conduct an internal audit, and declare itself in compliance. Self-declaration, however, is an approach that can lack substance and credibility.

2. The registration process consists of eight sequential steps, as follows: (1) decide to seek registration, (2) prepare internally, (3) determine EMS's functionality, (4) engage a registrar, (5) perform preliminary assessment and document review, (6) conduct formal audit, (7) eliminate nonconformances, and (8) secure registration.

3. When selecting a registrar, organizations should consider at least the following factors: past ethical performance, familiarity with the industry, reputation, and cost.

4. Environmental audits should be systematic, documented, objective, and conformance-oriented. In addition, the results are to be communicated to the client.

5. Environmental auditors should meet the following qualification criteria: at least a high school education; five years of work experience (four years if holding a college degree) in such areas as environmental science and technology, technical and environmental aspects of facility operation, environmental laws and regulations, environmental management systems, and audit procedures; formal training; on-the-job training; and personal competence, including speaking and writing skills, diplomacy and tact, objectivity, organizational skills, judgment, and sensitivity; and discretion. In addition to meeting all of these requirements, lead auditors must have more experience and demonstrate leadership capacity.

6. Three types of EMS audits are registration, surveillance, and internal. A registration audit is conducted by a registrar every three years. Surveillance audits are conducted by a registrar every six to twelve months to verify continued conformance. Internal audits are conducted by the organization's employees.

7. Objectives of the EMS audit include the following: confirmation of implementation by an independent third party, image enhancement among important stakeholders

and interested parties, and confirmation that the organization has a sound environmental foundation.

8. The audit scope establishes the boundaries of the audit. It specifically establishes the following: physical locations to be visited; organizational activities to be audited; resources required for the audit (audit team, organization's audit support requirements, and physical resources); and the manner of reporting on the audit.

9. The ISO 14000 audit attempts to determine the following: adequacy of the EMS documentation prepared in response to the requirements set forth in ISO 14001 and effectiveness of the EMS implementation.

10. The formal audit report is prepared by the lead auditor. Anything contained in this report should have been presented during the exit meeting. The formal audit report is a confidential document.

11. Following an audit the registrar can award registration, award conditional registration, or withhold registration.

## KEY CONCEPTS

| | |
|---|---|
| Approval | Environmental audit team |
| Audit process | Exit meeting |
| Audit report | Facility tour |
| Audit scope | Internal audit |
| Auditor qualifications | Major nonconformance |
| Certification/registration | Opening meeting |
| Conditional approval | Preliminary assessment visit |
| Corrective action | Registration audit |
| Disapproval | Registration process |
| End-of-day review | Self-declaration |
| Environmental audit | Surveillance audit |

## REVIEW QUESTIONS

1. What are the differences between certification/registration and self-declaration?
2. Describe the various steps in the registration process.
3. Explain the factors that should be considered when selecting a registrar.
4. Define the concept of EMS audit.
5. List and explain the various types of EMS audits.
6. What are the objectives of EMS audits?
7. What does the audit scope establish?

8. Describe the process that occurs when taking corrective action.

9. Explain the three options available to the registrar, in terms of registration, once an audit report has been completed.

10. Describe the process for preparing for the next audit.

## CRITICAL-THINKING ACTIVITIES

The following activities may be assigned as individual, group, or discussion activities to be completed in class or out of class.

1. Your company has decided to implement ISO 14001. The CEO is considering self-declaration, but you are an advocate of certification/registration. What can you say to persuade the CEO to support the certification/registration approach?

2. Complete the process of selecting a registrar for a manufacturing firm that employs 800 people. Document all of your activities and your ultimate selection.

3. Play the role of a registrar, and conduct a preliminary assessment and document review of a company that is willing to work with you.

## DISCUSSION CASE

ABC Paper Company, a paper processing company, has been located in the same community for more than fifty years. In the days before federal regulations, this company regularly discharged pollutants into the air and effluents into the local bay. Local residents referred to the river that transported the effluents to the bay as "Stink Creek."

Federal regulations forced the company to install scrubbers, stop discharging effluents, and implement several other pollution-prevention strategies. As time went by, "Stink Creek" was restored, the bay was cleaned up, and air pollution was minimized.

Now that the company has made such commendable progress, it is petitioning for permission to release the liquid by-products of the paper-making process into another bay. Scientists hired by the company are studying the issue and claim that these releases will not adversely affect the bay. Citizens claim that the company's track record in other states suggests otherwise.

### Discussion Questions

1. Is a role for ISO 14000 applicable here?

2. If the company were ISO 14000-certified, would its case appear more favorable to the community?

## ENDNOTES

1. ANSI/ISO 14010-1996, *Guidelines for Environmental Auditing—General Principles*, Clause 2.9, p. 1.

2. ANSI/ISO 14011-1996, *Guidelines for Environmental Auditing—Audit Procedures—Auditing of Environmental Management Systems*, Clause 3.3, p. 1.

3. ANSI/ISO 14012-1996, *Guidelines for Environmental Auditing—Qualification Criteria for Environmental Auditors*, Clause 7, p. 2.

4. ANSI/ISO 14010-1996, *Guidelines for Environmental Auditing—General Principles*, Clause 4.3, p. 2.

5. ANSI/ISO 14012-1996, *Guidelines for Environmental Auditing—Qualification Criteria for Environmental Auditors*, Introduction, p. v.

6. ANSI/ISO 14011-1996, *Guidelines for Environmental Auditing—Audit Procedures—Auditing of Environmental Management Systems*, Clause 4.2.3, p. 2.

7. ANSI/ISO 14001-1996, *Environmental Management Systems—Specification with Guidance for Use*, Clause 4.5.4, p. 5.

8. Ibid., Clause 4.6, p. 5.

9. Ibid., Clause 4.5.3, p. 5.

# CHAPTER SEVEN

# EMS Performance Improvement

## MAJOR TOPICS

- Improvement Versus Maintenance
- A Heads-Up for ISO 9000 Organizations
- Continual Improvement Requirements of ISO 14000
- Improvement Beyond Requirements
- What World-Class Organizations Do

## IMPROVEMENT VERSUS MAINTENANCE

It is important to have a common understanding of the terms *improvement* and *continual improvement* before reading this chapter. *Improvement* is often misinterpreted and misused, as shown in the following scenario.

A metal shop uses a chemical plating process. Eighteen months ago a spill occurred. The process was stopped quickly to prevent more of the chemical from overflowing. A team was activated to find the cause of the spill. It determined that an automatic valve had jammed in the open position, hence failing to cut off flow to the plating tank when the proper plating bath level had been achieved. A new valve was installed and tested with satisfactory results. The process was restarted and has experienced no spills to date. Has improvement taken place?

In another case, a fossil fuel-fired power generation plant uses "scrubbers" in its stacks to reduce emissions of combustion products into the atmosphere. The scrubbers are capable of removing 90 percent of the contaminants. During a routine check on emissions, it was found that only 50 percent of the contaminants were being removed. Plant technicians located three scrubbers that were not functioning effectively, and replaced them. Contaminant removal then went from 50 percent to 90 percent, as illustrated in Figure 7-1. Is this an improvement?

**Figure 7-1**
Restoring Historical
Performance—Maintenance

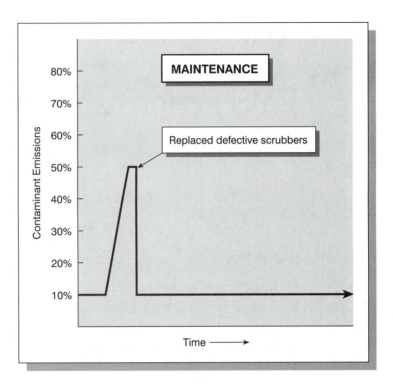

The answer to both questions is, *No, no improvement occurred.* In both cases the processes in question were only returned to their typical performance levels, which is called **maintenance**. Both of these processes can expect to experience the same problems in the future, since nothing was done to eliminate the root causes of the problems.

Suppose the power plant learned of a new technology that could enable it to lower stack emissions by another 2 percent. If the plant incorporated the new technology and it produced the anticipated results, would that represent an improvement? Yes, this would be an *improvement*, even though a small one. See Figure 7-2.

In the scrubber example, contaminant removal from stack emissions increased from 50 percent to 90 percent, which was called *maintenance*, not *improvement*. Then, when stack emissions were cut only 2 percent, this was called *improvement*. Why? Three points need to be clearly understood.

1. When a process's performance deteriorates and then is restored to its historic performance level, no improvement has occurred in the process's capability. The process has merely been returned to its normal performance (*maintenance*). Maintenance, however, is very important and is an essential element of any management system.

2. Although major, breakthrough improvements are wanted, any change for the better from a process's historic capability represents *improvement*—no matter how small. Most improvements fall into the "small" category. When the management system is targeting continuous improvement, small, incremental improvements result in sig-

**Figure 7-2**
Achieving Better Performance—
Improvement

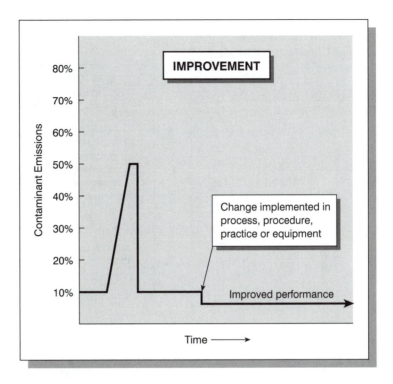

nificant improvement over time. That is the power of *continual improvement*. See Figure 7-3.

3. It is possible to achieve process improvement without changing the absolute performance of the process. If a process can be made more reliable or consistent, and therefore, less likely to fail, this is an improvement. Typically this kind of improvement is the result of locating and eliminating root causes, not just symptoms, of process variability or failure modes. Also, if a process can be made more difficult to be operated incorrectly—the Japanese call it **poka-yoke**, foolproofing the process[1]— this is an improvement. See Figure 7-4.

**Improvement**, therefore, takes the process to a new, higher level of performance or renders the process more reliable, more consistent, or less likely to permit operator-

---

**ISO 14000 INFO**

*"Putting out fires is not improvement. Finding a point out of control, finding the special cause and removing it, is only putting the process back where it was in the first place. It is not improvement of the process."[2]*
*Attributed to Dr. Joseph Juran by W. Edwards Deming*

**Figure 7-3**

Performance Improvement
Through Continual Improvement

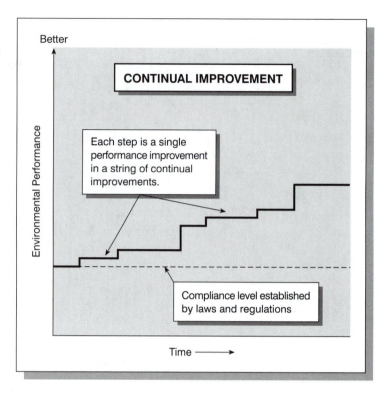

induced errors. **Continual improvement** is simply the relentless pursuit of process improvements on a continuing basis, never becoming satisfied with the current state. No matter how well a process performs or how reliable and consistent it is, it falls short of perfection, the real objective.

## A HEADS-UP FOR ISO 9000 ORGANIZATIONS

> This International Standard [ISO 14000] shares common management system principles with the ISO 9000 series of quality management standards. Organizations may elect to use an existing management system consistent with the ISO 9000 series as a basis for its environmental management system.[3]

This extract from the Introduction to ISO 14000 is an invitation for firms to use their **ISO 9000 quality management systems** (QMS) as the basis for their ISO 14000 environmental management systems. This seems to be a sensible approach to developing an EMS. However, organizations operating under ISO 9000 will need to take particular note of ISO 14000's position on *continual improvement*. The original version of ISO 9000 did not mention continual improvement. Further, the 1994 version makes only a brief reference to continual improvement in ISO 9000-1's Introduction and in ISO 9004-1, Clause 5.6. ISO 9000 does not require continual improvement. (*Note:* It is anticipated that con-

**Figure 7-4**
Improvement in Process
Reliability and/or Consistency

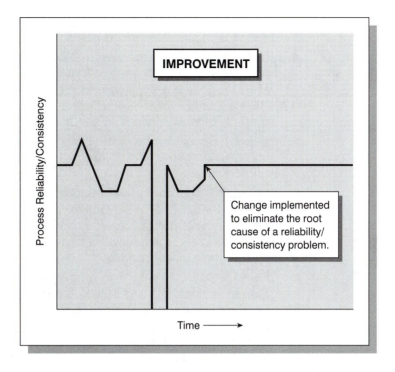

tinual improvement will be an important part of ISO 9000 when the Standard is released in its year 2000 edition.)

An organization accustomed to the demands of ISO 9000 will have to be doubly certain to incorporate continuous improvement into its EMS, especially if the ISO 9000 QMS is used as its basis. Continual improvement is a cornerstone requirement of ISO 14000.

## CONTINUAL IMPROVEMENT REQUIREMENTS OF ISO 14000

Continual improvement is an especially significant element of the ISO 14000 EMS, which is illustrated by the sheer number of references to it in the Standard, as explained next.

### ISO 14001

- Introduction, p. vii: "It should be noted that this International Standard does not establish absolute requirements for environmental performance beyond commitment, in the policy, to compliance with applicable legislation and regulations and to *continual improvement.*"

- Introduction, p. vii: Figure 1 shows *Continual improvement* to be the result of the ISO 14000 EMS model.

- Clause 3.1: Defines *continual improvement* as the "process of enhancing the EMS to achieve improvements in overall environmental performance in line with the organization's environmental policy."

■ Clause 4.2 b): Requires the organization's environmental policy to include "a commitment to continual improvement and prevention of pollution."

■ Clause 4.6: Requires top management to review at intervals the EMS "to ensure its continuing suitability, adequacy and effectiveness," and, as part of the review, "to address the possible need for changes in policy, objectives and other elements of the EMS, in light of EMS system audit results, changing circumstances and the *commitment to continual improvement.*"

■ Clause A.1, General Requirements: States in part, "The environmental management system provides a structured process for the achievement of *continual improvement. . . .*"

■ Clause A.2: States, "The environmental policy is the driver for implementing and improving the organization's environmental management system so that it can maintain and potentially improve its environmental performance. The policy should therefore reflect the commitment of top management to compliance with applicable laws and *continual improvement.*"

■ Clause A.6: States, "In order to maintain *continual improvement*, suitability and effectiveness of the EMS, and thereby its performance, the organization's management should review and evaluate the EMS at defined intervals."

■ In addition to these explicit references to continual improvement, the following clauses illustrate implicit references:

  • Clause 4.3.3, Objectives and Targets—established for *improvement*

  • Clause 4.3.4, Environmental Management Program(s)—established to manage *improvement* efforts

  • Clause 4.5.2, Nonconformance and Corrective and Preventive Action—the latter normally achieved through *continual improvement*

## ISO 14004

■ Introduction, Clause 0.1, Overview: Lists several key principles for implementing or enhancing an EMS, including, "Establish a management process to audit and review the EMS and to identify opportunities for *improvement* of the system and resulting environmental performance."

■ Clause 0.2: States that the EMS "can help an organization provide confidence to its interested parties that—the [environmental management] system's design incorporates the process of *continual improvement.*"

■ Clause 3.1: Offers the same definition of *continual improvement* as ISO 14001, Clause 3.1.

■ Clause 4: Lists the five principles of an EMS. Principle 5 states, "An organization should review and *continually improve* its EMS, with the objective of improving its overall environmental performance."

■ Figure 1: Shows *continual improvement* to be the result of the five principles listed in Clause 4.

- Clause 4.1.4: Lists *continual improvement* as one of the considerations and commitments of an environmental policy.
- Clause 4.5: Restates Principle 5 listed above.
- Clause 4.5.1: States, "A *continual improvement* process should be applied to an EMS to achieve overall improvement in environmental performance."
- Clause 4.5.3, Continual Improvement: States, "The concept of *continual improvement* is embodied in the EMS. It is achieved by continually evaluating the environmental performance of the EMS against its environmental policies, objectives and targets for the purpose of identifying opportunities for improvement."
- In addition to these explicit references to continual improvement, several other clauses make implicit references:
  - Clause 4.2.5, Environmental Objectives and Targets—established for improvement
  - Clause 4.2.6, Environmental Management Program(s)—established to manage improvement initiatives
  - Clause 4.4.3, Corrective and Preventive Action—the latter normally achieved through improvement to the process

Given the emphasis placed on continual improvement in ISO 14000, it is clear that the organization must incorporate the continual improvement philosophy as a key element of its EMS. Continual improvement in an EMS setting requires a relentless, never-ending cycle of improving the processes that have possible environmental impacts as well as improving the policies, procedures, and practices of the EMS. When problems arise, it is not enough to fix the problem; the processes or procedures must be improved to prevent recurrence of the problem. All relevant employees must constantly seek ways to improve processes, procedures, and practices to achieve improved environmental performance and to enhance process reliability.

Internal and external audits will have to consider the results of continual improvement. Records, therefore, must be kept as proof of improvement. Continual improvement results are measured in terms of improved environmental performance for processes with environmental impacts and through other relevant data.

## IMPROVEMENT BEYOND REQUIREMENTS

While ISO 14000 does not establish any environmental performance requirements, it does insist that the organization commit to compliance with all relevant environmental legislation and regulations. Therefore, in order to achieve registration to ISO 14000, the organization must, as a minimum, meet the legal and regulatory environmental performance requirements in effect in the community where the organization is located. ISO 14000 does not stop here; rather, it also requires that the organization commit to a program of continual **improvement beyond requirements**. If one starts at the environmental performance level that just meets the laws and regulations, and then applies continual improvement to the EMS and all its processes, procedures, and practices, then over time environmental performance will surpass what is required by the laws and regulations.

ISO 14000 requires conformance to the requirements of ISO 14001, which were detailed in chapter 3, and compliance with applicable laws and regulations and "other requirements to which the organization subscribes."[4] These "other requirements" may be licenses, permits, health and safety requirements, and other relevant requirements. Usually the organization has no choice but to comply with the instructions and restrictions of such requirements. However, there is another category of requirements included in the ISO catchall phrase, and that involves self-imposed requirements. The organization has control in this category.

Assume that an organization has a legal requirement of discharging no more than 200 parts per million (ppm) of some chemical. To be safe, the company designs its processes for discharging only 150 ppm; in other words, its procedures and practices establish a self-imposed limit of 150 ppm. If the process were to exceed the internal limit, this self-imposed limit would allow some latitude before causing problems with the EPA (or other relevant legal or regulatory agencies). The presumption is that internal alarms will signal when the 150-ppm threshold is crossed, giving the organization the opportunity to correct the problem before the legal limit is reached. The EPA will be satisfied, and the company will be happy. However, the ISO 14000 auditor will consider it a nonconformance if the 150 ppm internal limit is breached. Why? Because the organization established a limit of 150 ppm in the documentation that is part of the EMS. In so doing, they told the auditor what they were going to do (i.e., will not exceed 150 ppm). If the auditor finds that the process has reached, say, 160 ppm, this would be objective evidence of a nonconformance. The organization did not do what it said it would do. (*Note:* we have referred to this as a nonconformance, but in strictest terms, if the 150-ppm limit were part of the "other requirements to which the organization subscribes," it may easily be seen as a noncompliance, even though it is self-imposed.)

It is commendable if the organization wants to set its environmental performance limits tighter than those established by the legal and regulatory agencies, but the organization should use common sense in the establishment of internal limits and the creation of documentation relevant to them. In the example given earlier, it would have been possible to adopt the legal 200-ppm limit, and at the same time set procedures and processes to sound alarms at 150 ppm. With this approach, as long as the alarm sounds at 150 ppm there, is no nonconformance. Also, as long as the process is brought under control, or shut down, before the 200-ppm limit is reached, there is no noncompliance.

Many organizations consistently better legal and regulatory requirements, and it is anticipated that ISO 14000 will enable many more worldwide to do so. This is certainly an objective of ISO 14000. Nonetheless, organizations should guard against setting themselves up for needless findings of nonconformance or noncompliance by their registrars as a result of unnecessarily restrictive operating parameters.

There is another side to this issue. Limits that are within legal parameters, but are significantly less demanding than the process capability, may have a downside. Assume that an organization implements technology which results in cutting emissions to 20 percent of the relevant legal and regulatory requirements. Should the organization impose on itself a limit stricter than the legal one? If so, should it set the limit all the way down to 20 percent of the legal requirement, the performance level of which its process is now capable? Probably a stricter self-imposed limit should be established, initially at a level somewhere between the legal limit and the process's capability. Why is this recommended when ISO will enforce it even though it is not *legally* required? It is recommended because the approach ensures the organization that the new, clean process will not be allowed to deteriorate to a performance level significantly less than its capability. After the new process has proven itself to be reliable at the limit set, the limit should be reduced, then reduced again, until finally it is near the capability of the process. To do otherwise—allowing the process to deteriorate from its optimum capability so long as it does not exceed the legal limit—because there is no pressure to maintain it more closely, would be the equivalent of throwing the cost of the technology away. In other words, why spend the money for the new technology in the first place?

## WHAT WORLD-CLASS ORGANIZATIONS DO

**World-class organizations**, wherever possible, establish their internal environmental limits to tolerances stricter than required by laws and regulations. At the same time, they avoid needless nonconformances and noncompliances that can result from overly aggressive internal performance requirements.

World-class organizations always strive to eliminate pollution. One of the techniques through which organizations achieve this is continual improvement. Continual improvement can result in improved and cleaner processes, improved technology, and improved procedures and practices. As an integral part of their continual improvement process, they employ the PDCA Cycle, as described in chapter 3. Everyone in the organization—from top management to hands-on employees—is involved in searching for opportunities for improvement and ways, however small or grand, to accomplish it.

World-class organizations believe that "good enough" is never good enough. They constantly seek to raise the environmental performance bar.

═════════ SUMMARY ═════════

1. Extinguishing fires, plugging leaks, and returning performance to levels that preceded a problem do not represent *improvement*. This is *maintenance* and it plays a critical role in an organization's performance. However, improvement takes performance to a higher level or renders a process more reliable, more consistent, or less likely to permit operator error.

2. Organizations that have implemented ISO 9000 may use their existing quality management system (QMS) as the basis for their ISO 14000 environmental management system.

3. ISO 14000 emphasizes the need for continual improvement. In fact, the standard requires it. This means that an organization's environmental performance must get better and better incrementally forever. Consequently, just complying with legislation and regulations is not sufficient. An organization that complies today will have to improve on that performance tomorrow and continue forever.

4. World-class organizations typically establish their internal environmental benchmarks to tolerances stricter than required by laws and regulations. At the same time, they do not set the benchmarks unrealistically high.

## KEY CONCEPTS

Continual improvement                     Maintenance
Improvement                               *poka-yoke*
Improvement beyond requirements           World-class organizations
ISO 9000 quality management systems

## REVIEW QUESTIONS

1. Explain the difference between improvement and maintenance.
2. Develop an example that illustrates the difference between improvement and maintenance. Do not use the examples from your text.
3. Explain how a company that is certified to the ISO 9000 standard can save time in establishing an environmental management system for ISO 14000.
4. Develop an example that illustrates the concept of continual incremental improvement to environmental performance.
5. Can an organization that hopes to maintain its ISO 14000 certification just meet government regulations and legislation? Explain.
6. How do world-class organizations approach their preparation for ISO 14000 certification?

## CRITICAL-THINKING ACTIVITIES

The following activities may be assigned as individual, group, or discussion activities to be completed in class or out of class.

1. Why do you think ISO stresses so strongly the need for continual improvement? Develop a rationale justifying this focus on continual improvement.
2. If continual improvement is a lifelong process, organizations would be wise to chart their improvement goals over the long term. Select a government-mandated environmental regulation that would apply to either a manufacturing or a processing company. Establish the beginning benchmark as the required performance. Graph a ten-year projection for continual improvement.

## DISCUSSION CASE

Jones Processing Company had to invest hundreds of thousands of dollars in pollution-prevention equipment in order to comply with the federal Clean Air Act. Because there have been such drastic improvements with regard to airborne emissions, the company's CEO wants to pursue ISO 14000 certification. Her attitude is summarized best by the following statement she made during a meeting of top managers. "We've spent the money. We might as well have something to show for it." Unfortunately, she does not want to spend any additional money on pollution control unless absolutely mandated to do so by federal or state legislation. Her vice president for environmental and occupational safety is concerned about the ramifications of this attitude.

## Discussion Questions

1. What are the ramifications of the CEO's attitude toward future investments in pollution-prevention equipment?
2. How might the CEO's attitude affect her company's ability to earn and maintain ISO 14000 certification?

## ENDNOTES

1. Kiyoshi Suzaki, *The New Manufacturing Challenge—Techniques for Continuous Improvement* (New York: The Free Press, A Division of Macmillan, Inc., 1987), p. 98.
2. Mary Walton, *The Deming Management Method* (New York: The Putnam Publishing Group, 1986), p. 67.
3. ANSI/ISO 14001-1996, Introduction, p. viii.
4. Ibid., Clause 4.2 c), p. 2.

## CHAPTER EIGHT

# Implementing ISO 14000: Steps To Registration

## MAJOR TOPICS

- Organizational Decision to Implement ISO 14000
- Deciding Whether to Pursue ISO 14000 Registration
- Potential Registration Problems
- Minimizing Registration Costs
- Fifteen Steps to Registration
- Follow-Up to Registration

## ORGANIZATIONAL DECISION TO IMPLEMENT ISO 14000

The chairman of the U.S. Technical Advisory Group to ISO TC 207[1] notes that organizations are typically choosing to implement ISO 14000 for internal management system efficiencies, waste reduction, and to promote proactive compliance to regulations. A survey of ninety-nine firms lists the following specific incentives for registration:

1. Customer demand, 22%
2. Competitive advantage, 20%
3. Desire to improve environmental management system, 18%
4. Public relations impact, 9%
5. Influence government relations, 8%
6. Need to streamline existing programs, 6%
7. All others, 17%[2]

Many good reasons exist for implementing an ISO 14000 environmental management system. However, is the cost of implementing ISO 14000 justified for the organization's individual situation? Each organization will have to answer this question for itself, considering the following:

- Does the organization have environmental aspects as a result of its activities, products, or services?
- Does the organization risk the potential for fines or litigation as a result of environmental performance?
- Is there environmental pressure from customers, governmental agencies, shareholders, or outside groups?
- Is the organization's environmental standing with its community, industry, and government less than satisfactory?
- Is full compliance with applicable laws and regulations difficult for the organization?
- Do the organization's competitors have better environmental performance?
- Is the organization's environmental management system (with its processes and procedures) wasteful, inconsistent, or ineffective?
- Do competitors appear to have better control of environmental issues?
- Is the organization's environmental credibility an issue?
- Do the employees demonstrate an attitude inconsistent with sound environmental practices?
- Does the organization have difficulty obtaining environmental insurance?
- Can the organization afford the implementation initiative?

Without considering the last item for the moment, if the objective answer to any of the other questions is *yes*, then ISO 14000 implementation is probably in the organization's best interests. An EMS that conforms to ISO 14001 will help the organization eliminate any unfavorable responses to these questions.

With regard to the final question, developing an ISO 14000 EMS and its infrastructure will require money and time. However, before answering the question of whether it can afford such expenditures, the organization should carefully consider the points made in the remainder of this section. The real question may turn out to be, Can we afford *not* to make the investment?

The amount of money required can vary greatly depending on the nature and size of the organization. A small business can expect to spend approximately $10,000 to achieve registration. (Later in this chapter information will be provided on minimizing the costs of implementation and registration.) For most organizations the cost is not likely to be seen as a major factor.

The more difficult consideration involves the availability of employee time to work on the development of the EMS. As a result of the downsizing that has taken place in American business and industry, it is rare to find employees with free work time. More than considering how much free time is available to devote to initiatives such as ISO 14000 preparation, management needs to determine its priorities. Suppose an organization determines that it should implement ISO 14000. It may need to designate other tasks that can be slowed, or even stopped, in order to make time available for developing procedures, building an internal audit capability, and achieving the training that is needed. In other words, management must decide how to assign its employee resources for the long-term good of the organization.

Applying the lessons learned from ISO 9000, the following direct benefits will accompany the implementation of ISO 14000:

■ Better, more effective environmental documentation, including policies, procedures, practices, and records.

■ Positive cultural change resulting from employee training and their understanding that environmental issues are of vital concern to top management and the welfare of the organization.

■ Greater environmental awareness by all employees, especially those associated with activities having potential or real environmental impact.

■ Better environmental performance providing a wide range of collateral benefits, from public relations to avoidance of fines and litigation.

In chapter 1 we listed eleven more reasons for adopting the ISO 14000 EMS. Each organization will have its own reasons. ISO 14004 offers the following:

> An organization should implement an effective environmental management system in order to help protect human health and the environment from the potential impacts of its activities, products or services; and to assist in maintaining and improving the quality of the environment.[4]

As the world increases its awareness of the need, there is a worldwide trend to implement effective environmental management systems. Not too far in the future governments may establish ISO 14000 as a required EMS standard. Regardless of what future requirements might be, concerned organizations are incorporating ISO 14000 in their business strategies; having an effective environmental management system is simply good business. Adopting an ISO 14001 EMS is the right thing to do for both the strategic self-interests of the organization and for the benefit of the planet's ecosystem.

---

### ISO 14000 INFO

*ISO 14000 appears to be taking off even faster than did ISO 9000. In its first full year after being published in late 1996, ISO 14000 had 5,017 registrations in 55 countries. (At the end of 1998 the numbers increased to 7,887 registrations in 52 countries.) From a survey of some 130 organizations, primarily accreditation and certification bodies, " . . . the rapid take-up of ISO 14000 suggests that this new family of standards is . . . helpful to organizations in implementing environmental management systems which allow them to combine business efficiency with care of the environment."[3]*

## DECIDING WHETHER TO PURSUE ISO 14000 REGISTRATION

Following the decision to adopt the ISO 14000 EMS, an organization must ask, Should we pursue registration to the Standard? As ISO 14000 states, the Standard is intended for organizations that *self-declare* conformance as well as for those pursuing registration through accredited registrars.[5] Suppose an organization implements an EMS that conforms to ISO 14001, and the EMS is demonstrated to be effective and appropriate for the organization's activities. It even leads to continual improvement of environmental performance, which already complies with applicable laws and regulations. Should the organization self-declare conformance with ISO 14000, or are there overriding reasons to pursue the formal registration process?

Does registration promote even better environmental performance? This is a legitimate question, considering that our hypothetical EMS already conforms to ISO 14001 and is performing as intended. Registration *may not* improve the EMS or enhance environmental performance at all (although input from independent environmental auditors can be expected to be of great value for performance improvement). Registration *will* cost additional money. Management must determine if the additional cost is justified by the benefits.

Indeed, there are many significant benefits to be accrued through registration even though there may be no legal or customer requirement for it. Some benefits are:

- **Credibility.** Customers, legal and regulatory agencies, community officials, employees, and all other stakeholders and constituent groups will more readily accept (or trust) a declaration from an accredited registrar than they will from the organization itself.

- **Reduced exposure to liability and litigation.** An organization that is registered to the International Standard is in a much stronger legal position than one which may perform equally well but is not registered.

- **Accessibility to environmental insurance.** Environmental insurance is more readily available to registered organizations than those not registered.

- **Fewer audits.** Environmental audits by customers should be eliminated through ISO 14000 registration. Even legal and regulatory agency audits may be reduced or eliminated.

---

### ISO 14000 INFO

*Which of the following statements is more believable?*

1. *In a press release issued by the company today, ABC Chemical Corp. declared that it is in conformance with the International Environmental Standard, ISO 14000.*

2. *North American Registrars, an accredited certification body, has certified ABC Chemical Corp. as conforming with the International Environmental Standard, ISO 14000.*

■ **Continuing attention by top management.** Management has a tendency to put its attention in areas it enjoys most. This may not include environmental issues. However, when management knows that it must stay involved in the EMS or risk loosing registration, it will have greater incentive to stay focused on environmental performance.

Undoubtedly there will be organizations that choose to self-declare their ISO 14000 conformance with no intention of ever seeking registration, but the advantages of registration will lead most organizations to pursue it. However, there are organizations for which the best approach is to first implement the ISO 14000 EMS in the **self-declaration** mode, gaining experience and fine-tuning the EMS and its procedures and practices over a period of time. When the organization is satisfied that its EMS conforms fully to ISO 14001, is appropriate and effective, and that environmental performance is at least satisfactory, then the organization may involve a registrar to "make it official." The advantage to some organizations of approaching ISO 14000 in this way is the elimination of false starts and the necessary redoing of any nonconforming elements of the EMS while the registrar is engaged. If the organization has been serious, diligent, and objective in its EMS implementation, registration should be almost *anticlimactic*. This approach requires internal ISO 14000 expertise from the start. (This issue is discussed more fully later in this chapter under the heading Minimizing Registration Costs.)

Ultimately, every organization that implements a conforming EMS should take the necessary steps to achieve registration. Self-declaration does not provide the significant advantages of registration.

## POTENTIAL REGISTRATION PROBLEMS

Approximately 15 percent of firms seeking ISO registration fail to pass the **registration audit**. Some of the more likely deficiencies include the following:

■ Failure to address each of the ISO 14001 requirements. The organization must respond to every clause in Section 4.

■ Failure to integrate the concept of continual improvement into the EMS. There are only two absolute requirements in ISO 14000. One of them is continual improvement; the other is compliance with laws and regulations.

■ Failure to incorporate the five EMS principles.
  • Commitment and policy
  • Planning
  • Implementation
  • Measurement and evaluation
  • Review and improvement

■ Failure to adequately determine and list the environmental aspects and their significance.

■ Failure to implement a system of targets and objectives to deal with the environmental aspects.

- Inadequate performance monitoring, measurement, reporting, and review.
- Inadequate internal auditing.
- Failure to provide adequate control over all levels of documents and the data in them. (For example, there should be no handwritten and initialed changes unless a formal update is assured within a few days, not weeks or months.)
- Uncalibrated tools and gauges. The organization must be able to show that tools and gauges used for environmental monitoring and measurement provide true and consistent readings.
- Failure to adhere to published procedures and practices.
- Failure of employee credentials to match published job description requirements. New employees must receive the environmental training necessary for their assignments.
- Lack of periodic management reviews of the EMS or inadequate review records.
- Management responsibility breakdown, including
  - Failure to define responsibility and authority for personnel
  - Assigning responsibility without granting the necessary authority
  - Failure to carry out management reviews of the EMS to ensure its continued appropriateness and effectiveness

Some other common registration problems are:

- Misinterpretation of ISO 14001 requirements. This potential problem must be avoided or both time and money will be wasted.
- Overdoing the development of ISO 14000 documentation. Quantity is not necessarily quality. EMS documentation should contain only the policies, procedures, practices, and records that are *necessary*. It should be clear and concise. If a procedure has no bearing on environmental performance, it should be omitted from the EMS. Remember, if a peripheral procedure is inserted into the EMS documentation, it will

---

### ISO 14000 INFO

*One of the most common problems encountered by registrar auditors is failure to follow documented procedures. Such a failure is a sign of haphazard management. When documented procedures are not followed, one of two things is happening. First, procedures that are considered to represent the best practices—and, therefore, have been documented for the purpose of standardization—are being ignored. When procedures and practices are ignored, environmental accidents happen. The second possibility is that some employees have discovered better procedures than those in the documentation and are using them to improve process performance and/or consistency. If this is the case, the new procedures should be documented, standardized, and used by all relevant employees.*

be subject to the same auditing attention as the critical procedures. The best idea is to start with a minimalist approach and expand documentation only as this becomes necessary.

■ Overcontrolling through the EMS. The word that best describes this problem is *overkill* and is typified by requiring multiple chain-of-command approvals for every action. Organizations should guard against applying so much control that the system becomes a burden rather than a benefit. Too much control inhibits improvement activities by making change too cumbersome. Remember, the EMS should be designed to work *for*, not against, the organization. If the EMS is burdened by overkill, the auditors will enforce it. The best approach is to control what must be controlled, but only to the degree necessary. Control can be increased later if necessary.

It is interesting to note that, technically, at least, it is possible to achieve registration while being out of compliance with legal or regulatory requirements. The organization, not the registrar, determines which laws and regulations are applicable to the organization's activities.[6] A law or regulation could be overlooked and not included as part of the compliance requirements of the EMS. In such a case the organization could be noncompliant with a law or regulation without the registrar's knowledge. In reality, the auditors (who should have knowledge of the laws and regulations applicable to the organization's industry) would probably advise their client even though they are not required to do so; failure to do so could lead to embarrassment for the registrar.

## MINIMIZING REGISTRATION COSTS

When considering the cost of registration, look at the whole scheme. Internal costs for developing the EMS, its documentation, and training must be combined with external costs (consultants, the registrar) to determine the true cost of ISO 14000 registration. In fact, the cost continues further. Each time the registrar conducts a follow-up or a surveillance audit, additional costs are incurred. Combine all costs to attain an accurate view of the true cost of ISO 14000 registration and operation. Most organizations will want to minimize these costs while realizing the maximum benefit from operating as an ISO 14000 firm.

---

### ISO 14000 INFO

*ISO has been careful to avoid placing registrars in legal jeopardy. Consider what might happen if a registrar were responsible for ensuring that all the laws and regulations applicable to a client's operations and activities were incorporated in the EMS. If something were overlooked, the registrar could conceivably be held legally responsible for any resulting noncompliance. In order to prevent this from happening, registrars are not given this responsibility. The client organization must determine which laws and regulations are applicable.*

> ### ISO 14000 INFO
>
> *ISO 14000 as an investment:*
> *In a survey of 500 ISO 14000-registered companies, 80% found the environmental management system to be cost effective. Sixty percent of the firms found the payback period for their EMS investment to be less than 12 months.[7]*

Considerations for minimizing costs include the following:

■   Are consultants needed?

■   Should the EMS be developed in-house or out-of-house?

■   Should the documentation be developed in-house or out-of-house?

■   Which registrar should be chosen?

■   What is the timetable?

Each of these topics is discussed below.

## Consultants

Many firms achieve registration without the help of outside **consultants**, and more could do so if they tried. Perhaps the best consultant is someone who knows your products or services and is familiar with your processes, culture, organization, and its personnel. Such a person, not surprisingly, usually is an employee. The right employee trained in ISO 14000 will be far more effective than the ISO 14000 expert who is unfamiliar with the organization or its business. If you pay consultants long enough to learn what they need to know about your organization, registration costs can skyrocket.

 Where do you find this employee who is also trained in ISO 14000? You select one and send him or her to one of the ANSI/RAB-accredited ISO 14000 lead auditor courses. The course lasts five work days and costs approximately $1,500. With this training, the employee will be able to perform as your in-house consultant in preparing for registration. Realize, however, that time spent in this capacity still represents a cost of registration, although a cost that will be well below the cost of an external consultant.

## Developing the EMS In-House or Out-of-House

An outside consultant can develop an EMS. Yet the results may not be acceptable, simply because an outsider cannot fully understand the organization and its products or services, vision, culture, processes, and other operational factors. If a consultant stays on-site long enough to learn about these things, it will probably cost more than the alternative of doing the job in-house. Actually, determination of **developing the EMS in-house or out-of-house** should be based on just one criterion: What is the best way? It is always better to have the people who will use the EMS develop it, meaning the organization's top managers. Nobody knows the organization, its products and services, and its culture better. Nobody knows better where the organization is trying to go or what it wants to become. This is critical information in the development of a viable EMS. In-house employees may not have an adequate understanding of ISO 14000 requirements, but these requirements can be learned. It is recommended that the organization develop the EMS in-house regardless of the cost factors. In the long run this approach will cost less anyway—and it will ensure that the organization rather than an outsider benefits from the learning that occurs.

## Developing Documentation In-House or Out-of-House

The consideration of **developing documentation in-house or out-of-house** is similar to that presented in the previous section, except that the creation of documentation should involve the process owners. At this level it will be necessary to understand the organization's processes, the gaps between the current documentation status and that required for ISO 14000, and what must be done to close the gaps. These are excellent tasks for teams of six to eight people including those directly involved with the processes. Again, regardless of cost considerations, it is wise to develop documentation in-house; any other way is likely to produce unsatisfactory results. The in-house approach will be less expensive in the long run, and it will ensure that the benefits of learning accrue to the organization.

Organizations already registered to ISO 9000 (or who are well along toward achieving ISO 9000 registration) have a definite advantage in developing the ISO 14000 EMS and its documentation. These organizations will have learned by doing, and the same techniques, practices, and even document formats can be shared. Certain things already in place for ISO 9000 can be used directly for ISO 14000; some may require minor modification (for example, organizational structure, training procedures, policies). ISO 9000 organizations need to look at ways to use what they have learned and to maximize sharing between the two standards. This will save both time and money.

## Choosing a Registrar

One of the considerations for selection of a registrar should be cost. Registrars have their own rate structures that can vary widely. In addition to rates, travel expenses are a major cost element. An organization should narrow the selection to registrars who are familiar with its industry, who routinely work with the same size of business, who enjoy a good reputation, and who are located in the general geographic area. Solicit bids from at least three registrars. Ask the price of the registration audit and the preliminary work that is a part of the initial registration process, a follow-up audit (in case the organization does not pass the first time), and surveillance audits.

## Registration Timetable

How long should it take to secure registration? What is your preferred deadline? A general rule states that the longer something takes, the more it will cost. This rule may not apply to ISO 14000 registration—at least not if the time frame is within reason. It would be helpful to do as much as possible in-house without the assistance of consultants. Remember, however, that employees still have their normal duties to perform and cannot spend full-time on ISO 14000 preparation. This tends to extend the **registration timetable**. If a longer schedule is acceptable, say eighteen months rather than one year desired, proceed with the in-house approach. If not, contract for outside help and accept the corresponding costs. Also, remember that every hour your employees spend on ISO 14000 work is part of the cost of registration. However, such costs can be recouped through better performance as employees come to understand their processes more fully as a result of being involved in the development of the EMS.

# FIFTEEN STEPS TO REGISTRATION

The following model applies to organizations seeking ISO 14000 registration and organizations planning to operate an ISO 14000 EMS without registering. Bracketed [ ] text indicates differences for organizations not seeking registration. Figure 8-1 also cues these differences.

Certain steps must be taken in preparation for ISO 14000 registration, and several steps have an order that should be observed. For example, individuals in the organization who see the need for registration would be foolish to start the preparation before securing the backing of top managers. Implementing an ISO 14000 EMS and securing registration is not an undertaking a group of enthusiasts can do without the knowledge and support of top management, since there will be significant costs involved and a major investment of employees' time. Similarly, even with the support of top management, it would not be prudent to start changing procedures and publishing an EMS manual before determining the organization's current posture in relation to the requirements of the Standard. The fifteen steps to registration, taken together, should be considered a model that may be adapted to the particular circumstances and needs of individual organizations. Refer to Figure 8-1.

*Note*: The fifteen steps described here are also represented in less detail in the flow diagram of Figure 6-1, ISO 14000 Registration Process. However, the element numbers of Figure 6-1 and the step numbers in Figure 8-1 do not correspond and are not related.

## Step 1: Secure Commitment from the Top

Commitment to achieving registration [to implementing a conforming EMS for organizations not seeking registration] from the highest level of management in the organization is an essential prerequisite to starting the ISO 14000 journey. Several primary reasons support securing **commitment at the top**, including the following:

■ The need for resources that only top management can authorize. These resources include money and employee time, both in significant amounts.

■ Inevitably there will be individuals who prefer the *status quo*. When they hold senior positions in the organization, it takes intervention by top management to overcome their resistance.

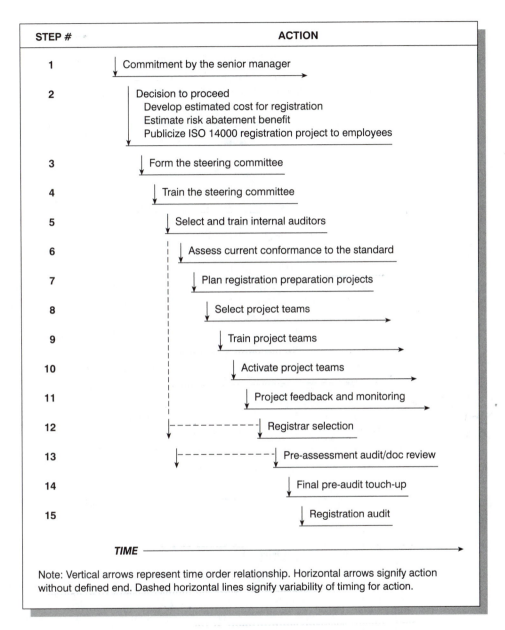

| STEP # | ACTION |
|--------|--------|
| 1 | Commitment by the senior manager |
| 2 | Decision to proceed<br>　Develop estimated cost for registration<br>　Estimate risk abatement benefit<br>　Publicize ISO 14000 registration project to employees |
| 3 | Form the steering committee |
| 4 | Train the steering committee |
| 5 | Select and train internal auditors |
| 6 | Assess current conformance to the standard |
| 7 | Plan registration preparation projects |
| 8 | Select project teams |
| 9 | Train project teams |
| 10 | Activate project teams |
| 11 | Project feedback and monitoring |
| 12 | Registrar selection |
| 13 | Pre-assessment audit/doc review |
| 14 | Final pre-audit touch-up |
| 15 | Registration audit |

TIME

Note: Vertical arrows represent time order relationship. Horizontal arrows signify action without defined end. Dashed horizontal lines signify variability of timing for action.

**Figure 8-1**
Fifteen Steps to ISO 14000 Registration

- The most important aspect of *leadership* is setting a positive example for employees to follow. In order for the ISO 14000 EMS to gain acceptance in an organization, top management has to demonstrate its commitment by being actively and visibly involved in the preparation process and it must be supportive of protecting the environment. Only then will employees commit themselves.

No matter how enthusiastic middle or even senior managers are, they cannot overcome a lack of commitment from the top.

## Step 2: Decision to Proceed

Having secured the necessary commitment from the top, the **decision to proceed** follows naturally. Even so, it helps to view it as a separate step in which certain important tasks occur. These tasks include the following:

- Develop a rough estimate of the costs for external services and time for internal tasks. At this point the estimate cannot be very precise because there are many things still unknown (primarily the scope of the internal work required), but a range can be developed within which the actual costs will likely fall. By estimating in this way, management can gain a sufficient grasp of the total cost in order to make an informed decision whether or not to proceed.
- An estimate of risk abatement should be included in the decision-making process. All organizations having environmental aspects associated with their activities, products, and services have a potential risk of environmental accidents that could lead to expensive legal problems. The organization should compare its current risk situation to that which would exist with an ISO 14000 EMS in place.
- With the decision to proceed made, management should begin publicizing the ISO 14000 effort to all levels of employees. When communicating with employees, managers should let them know that their input and support will be required, and should attempt to enhance their environmental awareness. Employees need to know what is going on.

## Step 3: Form a Steering Committee

After deciding that the potential investment for pursuing ISO 14000 registration [or implementation of an ISO 14000 EMS] is not a roadblock, and after having begun acquainting employees with the project, the next step is to choose an approach for managing and distributing the work of preparing for registration [or implementing the EMS]. A number of approaches can be used to manage the project. One frequently used, and seemingly promoted by ISO 14000, is to designate a Management Representative, who leads and coordinates the establishment, implementation, and maintenance of the EMS. ISO 14001, Clause 4.4.1, states:

> The organization's top management shall appoint (a)n specific management representative(s) who, irrespective of other responsibilities, shall have defined roles, responsibilities and authority for

a) ensuring that environmental management system requirements are established, implemented and maintained in accordance with this International Standard;

b) reporting on the performance of the environmental management system to top management for review and as a basis for improvement of the environmental management system.

From the standpoint of good management practices and the Total Quality Management philosophy, this is a flawed concept. Authority and responsibility should not be delegated to any individual(s) for the establishment (development) or implementation of the EMS. The preferred approach is to use a **steering committee** and project teams. Experience with Total Quality Management implementations has shown that the senior manager must retain the responsibility for developing and implementing any EMS. It is a fundamental function of leadership and cannot be delegated.

However, CEOs are not in a position to do it themselves. Consequently, the best course is to establish the senior management team as a steering committee chaired by the top manager (CEO, president, or another title). The steering committee provides leadership in establishing and implementing the EMS and in monitoring its performance. This includes all of the work necessary for registration. However, it does not mean, as an example, that new procedures must be written by the steering committee itself. Rather, the steering committee's role is to determine what is needed, secure the resources to satisfy those needs, and manage the activities of those given assignments. In the case of procedure writing, the steering committee would assign employee teams to accomplish the tasks and then monitor their progress.

ISO 14000 allows this approach. Although some registrars might abide by a literal interpretation of the cited clause, most would not. In the worst case, the top manager could designate himself as the Management Representative, or alternatively, designate some or all of the steering committee as management representatives. Either of these cases would satisfy even the most literal registrar interpretation. (ISO 14000 allows multiple management representatives, whereas ISO 9000 does not.)

## Step 4: Steering Committee Training

At this point some members of the senior management staff (the steering committee) still are uninformed concerning ISO 14000. Therefore, it will be necessary to provide **steering committee training** that covers at least the following topics:

- Familiarity with the ISO 14000 Standard and its guidelines
- The environmental philosophy contained in ISO 14000
- ISO 14001 requirements
- Applicable legal and regulatory requirements and any others to which the organization subscribes
- The work likely to be necessary to satisfy the ISO 14001 requirements

The steering committee members also must understand the rationale for undertaking the registration project [or implementation of the ISO 14000 EMS], that the work does not end with registration [or with implementation], and that ISO 14000 will be a normal part of doing business forever. If teamwork is not already an integral part of the organization's culture, a team-building seminar should be included in the steering committee's training since members will have to work effectively as a team.

With the steering committee approach, all senior managers automatically assume roles in the ISO 14000 registration process [or in the EMS implementation]. This may be the most powerful argument in favor of the steering committee approach. Any other approach will involve designating an employee as registration project leader [or as EMS implementation leader]. Typically a designated project leader will not have the authority or organizational standing to effectively coordinate and direct activities throughout the organization. When these activities require the time of employees, as they will, the project leader will probably have a difficult time with department heads. The steering committee approach eliminates this problem, because department heads and their employees understand that the issue in question is important to the "boss" and, as a result, are more likely to get the job done.

## Step 5: Select and Train Internal Auditors

Although it is too early at this point for internal auditors to begin checking the performance of the EMS, it is not too early for **internal auditor selection and training**. The sooner they are trained, the sooner their knowledge of ISO 14000 will begin paying off. That is why they should be selected and trained in this step. The number of internal auditors needed will depend on the size and complexity of the organization. The required minimum number can be no fewer than two, since an auditor must always be independent of the management of the activity being audited (see ISO 14001, Clause A.5.4). With just one auditor, there will always be one function in the organization that cannot be audited due to the impartiality rule. The actual number of internal auditors will have to be determined by the organization. In most cases, the auditing function is not a full-time job. Rather, it consists of specific audit tasks that must be accomplished periodically. It is a good idea to have enough internal auditors to allow audits to be conducted according to the required schedule, unencumbered by workload fluctuations.

A restraining factor is the cost of training. To ensure effectiveness, the internal auditors should complete an ANSI/RAB-accredited lead environmental auditor course. The course costs approximately $1,500 per person, and the duration is typically five days. If travel is required in order to attend, travel and per diem costs must also be considered.

By designating and training the internal auditors at this point, the organization can gain the benefit of their ISO 14000 knowledge throughout the registration [or throughout the EMS implementation] process. Having internal auditors during the preparation process is like having your own in-house consultants, only less expensive.

*Note*: Figure 8-1 shows the selection of the registrar as Step 12, which is the latest possible point for this task. It could occur as early as between Steps 5 and 6 since by this time the steering committee should be capable of making an informed selection. If the organization intends to use outside help for its preparation work, an early selection of the registrar could be advantageous. For example, the organization could have the registrar perform an early preassessment visit to determine where preparation efforts should be focused. This would replace Step 6. Some organizations, on the other hand, will want to become better prepared before hosting a registrar visit. These organizations will select their registrar at Step 12 and proceed from Step 6 as presented in Figure 8-1.

## Step 6: Assess Current Conformance

Whether this step is performed by an outside party or an in-house team, it must be completed before serious preparation work is started. Failure to observe this recommendation can result in wasted effort and cost. An effective assessment of the organization's current level of conformance to ISO 14001 and compliance with legal and regulatory requirements can be accomplished using in-house personnel. The internal auditors, if they have been trained, have the necessary expertise for this task.

The task involves taking an objective look at all relevant activities of the organization, with an emphasis on the present EMS (if any) and whatever activities, products, services, or processes relate to the environment. The organization's environmental aspects will be defined here. The goal is to determine what needs to be done to satisfy the requirements of ISO 14001 and applicable laws and regulations. To assist in this task, the ISO 14000 Checklist (see chapter 10) can be used. This instrument is easy to use, and its questions are linked directly to the requirements of ISO 14001.

By completing an assessment that compares the organization's current situation with the requirements of the Standard, the organization will learn where to focus its preparation efforts. The results of this initial assessment should not be assumed; ultimately the organization must conform to the satisfaction of the registrar [if the organization is registering]. There is no point in proceeding without a thorough understanding of current strengths and weaknesses relative to the Standard.

If the organization does not have the in-house expertise to conduct a valid assessment, it should have a consultant do it. In either case, the assessment must be done before proceeding to Step 7.

## Step 7: Plan Preparation Projects

Using the results of the initial assessment, develop a list of tasks that must be performed to bring the organization into conformance; in other words, **plan preparation projects**. The steering committee should convert this list of tasks into a timeline chart. A timeline is important because some tasks must be completed before starting others. For example, if one task is to develop an environmental policy, this task will have to be completed before other parts of the EMS manual can be developed, because everything must support the policy. Likewise, process flow diagrams may be required before process instructions or work instructions can be written or verified. The timeline chart is also necessary for the steering committee to plan a schedule for the registration [or EMS implementation] project. The registration date may be dictated by customer demand (e.g., terms of a contract), or it may be a target that the steering committee believes is reasonable. Either way, the timeline chart of tasks should be set up to accommodate the completion date of the registration [or implementation] project. The timeline represents the plan for tasks to be accomplished and the points in time when they must be completed. To start a project of this magnitude without a plan would virtually guarantee failure.

## Step 8: Select Project Teams

Once the plan has been developed, the steering committee then determines the composition of the teams that will be assigned specific tasks. **Project team selection** is more involved than one might think. For example, if a task is to develop an understanding of a particular process and then document the process, there will be some "givens."

1. Always put the process owner(s) on the team. The owners live with the process and its idiosyncrasies on a daily basis; they are, therefore, best able to document how the process really works and what they need to do to make the process do its intended job, or do it better.
2. Include the supplier and customer of the process—that is, the operators of the preceding and succeeding processes—on the team. They will have invaluable insights concerning the process.
3. Ensure that the team is cross-functional. It will need someone with unbiased views—an outsider—to ask questions or to make suggestions that someone more familiar with the process might overlook or might not ask. Include other functions that have a direct input into the process.

Beyond this, the steering committee is advised to select team members with the proper attitude to support the activity. The members ought to be positive toward both team activities and change. The work will be difficult enough without negative, contrary team members.

## Step 9: Train Project Teams

Employees of many organizations beginning the ISO 14000 journey will be untrained in some essential subjects. This does not apply to organizations operating according to the total quality philosophy; but for most other organizations it will be necessary to provide new teams with **team training** in the following areas before activating projects.

- *ISO 14000 and Environmental Management.* (This item also applies to total quality organizations.) Team members need to understand what ISO 14000 is all about and that it is ultimately tied to protection of the environment and sustainability relative to using natural resources. They should also learn why top management has concluded that ISO 14000 registration [or EMS implementation] is vital for the organization. All team members need to have this understanding; without it, the team cannot achieve the focus necessary to carry out the tasks to be completed. If the internal auditors have been trained, they will be able to conduct this training, along with a senior manager who can speak to the rationale for seeking registration [or EMS implementation]. This session can be covered in approximately two hours.

- *Team-Building Training.* In most organizations it will be necessary to provide training on the subject of how individuals can work effectively together as a team. Remember, this may be a new experience for many employees, even those who think they are good team players. There are a number of books on this subject, including Peter Sholtes' excellent work, *The Team Handbook*,[8] and Goetsch and Davis' *Quality Management: Introduction to Total Quality Management for Production, Processing, and Services.*[9] The organization may need to hire an outside consultant to conduct this training if no one on the staff is qualified. As an alternative, an employee skilled in conducting training classes might be sent to one of the many available seminars in order to develop the necessary expertise. Having an in-house trainer is advantageous since the training will have to be repeated as new teams are formed. Using an outside consultant can be expensive. This session can be completed in approximately four hours.

- *Use of Quality Tools.* Before a new team can be given the go-ahead on a project, it must be equipped with the tools necessary to accomplish the task. For many of the ISO 14000 projects, the necessary tools are the same as those used in a Total Quality Management environment. TQM uses nine fundamental problem-solving and analysis tools. They are:

  - *Pareto Charts*: To separate the significant from the less important.
  - *Cause-and-Effect (Fishbone) Diagrams*: To identify and isolate root causes of problems.
  - *Stratification*: Grouping data by common elements or characteristics to facilitate interpretation.
  - *Checksheets*: To facilitate collection and interpretation of data.
  - *Histograms*: To depict the frequency of some occurrence within a process.
  - *Scatter Diagrams*: To determine the correlation between two variables.
  - *Run Charts and Control Charts*: To record the output of a process over time and to separate *special* and *common* causes of variation, respectively.
  - *Flow Charts*: To describe inputs, steps, functions, and outflows so that a process can be understood, analyzed, and improved.
  - *Surveys*: To obtain relevant information from sources that might not otherwise be heard from.

All of these tools will at one time or another be useful to members of the ISO 14000 project teams, but a core set of tools should be taught to all of the newly formed teams. That core set should include flow charting, Pareto charts, cause-and-effect diagrams, surveys, and stratification. For teams tasked with determining the effectiveness of processes, the additional tools of run charts and control charts, histograms, check sheets, and scatter diagrams should be added. Assuming that the teams will have a facilitator available who is competent in applying the tools (probably the same person who does the tools training), the core set can be taught in four hours. Then the facilitator can assist team members one-on-one.

■ *Timing of the Training.* Training is not retained unless it is applied. The sooner it is applied, the better the retention. The best approach is to train the teams just as they are activated. One can make a case for training larger groups all at once. Initial direct training costs are certainly lower with this approach, but the true cost of training must also take into account the long-term effectiveness of the training. If all employees are trained in one large session, but are unable to immediately apply what they have learned, the information will be forgotten and the training will, as a result, be rendered ineffective. Ineffective training, no matter how low the cost, can be more costly in the long run than effective training, even though the initial costs are more.

## Step 10: Activate Project Teams

The steering committee has identified a preparation project and identified employees to serve on the team for this project. The team has received the training it needs to accomplish the task. The temptation at this point is to simply tell the team to start working. However, experience indicates that **project team activation** should be a formal, structured, interactive process in which the team is given a written charter that documents exactly what is expected. In the absence of such a charter, the team members may be unclear about what they should do and how much authority they have. The team activation meeting is used by the steering committee to do several things.

■ First, the team should be given an overview of the project, an explanation of what led to the development of the project, and a description of any problems or issues perceived by the steering committee to be relevant to the project. The team should also be informed as to why its particular members were selected and what the steering committee expects of them.

■ Second, the schedule of meetings between the team and the steering committee should be set. Through this vehicle team members will know how often they should report, to whom, and in what format. This schedule will ensure a formal closed-loop system (operating in the PDCA Cycle) for two-way feedback and interaction between the team and the steering committee.

■ Third, the steering committee should give the team a proposed schedule for completion of individual tasks and the overall project. In Step 7, the steering committee planned for the overall registration [or for the EMS implementation] project and placed the individual tasks on a timeline. Now this information is imparted to the team.

■ Fourth, the steering committee must assign the team responsibility for accomplishing the project in question, define the authority of the team, and ensure that team members understand both their responsibility and their authority. With these things understood, team members will know what they can do on their own, as well as when they must ask the steering committee for approval or assistance. In the absence of a clear understanding of responsibility and authority, teams will flounder, especially when confronted with cross-functional considerations.

In addition, the team should be asked to select a team leader. In some cases the team leader may be designated by the steering committee, but it is usually better to have the team make its own selection. The team leader's responsibilities are to conduct meetings, assign action items to team members, and communicate as necessary with the steering committee between formally scheduled meetings. The leader is also responsible for having a team member record minutes of meetings, including action item assignments. The leader should publish the agenda for all team meetings, distributing them to team members a day or two before meetings.

■ Fifth, the steering committee should make a trained facilitator available to the team. The facilitator should attend all meetings. His or her job is to keep meetings focused and moving in the right direction. The facilitator also provides assistance in the use of the appropriate tools (discussed in Step 9) and keeps discussions moving while, at the same time, preventing domination by individual team members. He or she also summarizes discussions, brings discussions to a close when appropriate, prompts discussion when necessary, and helps maintain the integrity of the agenda in terms of both time and content. As experience is gained, facilitation functions may be assigned to a team member, but an outside team facilitator can be helpful in the early stages of teamwork.

All of the items explained in this step should be thoroughly discussed at the team activation meeting and included in the written team charter.

## Step 11: Project Feedback and Monitoring

The **project feedback and monitoring** phase of the registration [or EMS implementation] preparation process begins with the activation of project teams in Step 10. For the duration of the team projects, the steering committee will receive feedback from all teams in accordance with the schedule and format agreed to during the activation meeting (also Step 10). The steering committee uses this information to monitor progress and to provide new instructions for the team as appropriate. Thus, a closed-loop PDCA Cycle exists between the steering committee and the teams.

## Step 12: Select a Registrar

[This step is not applicable to organizations not seeking formal ISO 14000 registration.]

Companies typically select their registrars six to eighteen months prior to the target date for the registration audit. As noted earlier, selection of the registrar could occur as early as between Steps 5 and 6. The choice is based on whether or not the organization wants to use the registrar to assist with some of the intervening steps. This step is the latest point at which **registrar selection** should occur. Refer to the section Choosing a Registrar under the heading Minimizing Registration Costs in this chapter.

When **selecting a registrar**, be sure to do the following:

■ Check with trade organizations and the accrediting bodies for background information on potential registrars. (In the United States the accrediting body for ISO 14000 is ANSI/RAB.)

- Solicit references on registrars from other companies, preferably those in your own industry.

- Question potential registrars directly. The general tone of responses may be as revealing as the answers themselves.

- Make sure that potential registrars can accommodate your schedule for registration.

Consider the following critical factors in the registrar selection process:

- Does the registrar have proper accreditation? Note that registrars in the United States are normally accredited through ANSI/RAB, but firms accredited by foreign agencies are certainly acceptable. A number of European registrars with European accreditation do business in the United States and Canada.

- Does the registrar have sufficient general experience as well as specific experience in your industry? You may not want to be among a new registrar's first clients, because its learning curve could be expensive for you. However, the more important issue is, Does the registrar speak the language of your industry and understand the kinds of processes you use? Avoid any situation where time is wasted educating your registrar.

- Does the registrar have the resources to satisfy your requirements? Your requirements include a registration audit, a preassessment visit, (possibly) follow-up visits, and surveillance audits. The registrar should be able to accommodate your schedule without requiring major changes to it.

- Does the registrar use its own full-time auditors, or does it engage independent auditors? Independent auditors are acceptable provided they are accredited by RAB. In fact, with independent auditors it can be easier to assemble an audit team that is knowledgeable in a particular industry. However, one significant problem exists with this approach. If the registrar uses independent auditors for its audits, it is likely that you will lose continuity as members of the audit team change from audit to audit. This is important because continuity from audit to audit is valuable. An audit team that is familiar with your organization does not have to waste time acclimating for every visit. Rather, it knows what to look for based on previous visits. You will benefit from the consistency of a stable team. In addition, costs should be lower since a stable team becomes familiar with all aspects of the organization and can complete its work faster.

- Have you compared all estimated costs from potential auditors (i.e., follow-up and surveillance audits, preassessment visits, and the registration audit itself)? Be certain that the registrars know the size of your firm. The bigger the firm, the higher the registrar's costs, since more auditors will be required and for a longer time.

- Is the "chemistry" between the registrar and the organization favorable? You will use the selected registrar for several years, so it is important to choose one with whom you are comfortable.

- What is the reputation of the registrar? Indeed, this may be the most important factor in registrar selection. The ultimate viability of your registration depends on the reputation of your registrar. *Never* select a registrar because it has a reputation for being "easy." Organizations that make this mistake never get the full benefit of ISO

14000, for eventually they are labeled in the marketplace and by regulatory agencies. You need—and ISO 14000 needs—a registrar that puts ethics at the top of its priorities, followed closely by competence and the willingness to work with clients to ensure valid, fair, and equitable audits. Put as much emphasis on reputation and ethics as you put on cost. Selecting a good registrar is like selecting a good college: You may have to work harder and longer, but you will learn more and be better prepared for the future if you are educated at a top-ranked college.

## Step 13: Preliminary Assessment Audit and Document Review

Although not required, most firms contract for a preliminary assessment audit. They hope that the **preliminary assessment audit and document review** will help them prepare for the registration audit. The preliminary assessment audit is conducted to identify ISO 14000–related deficiencies that have to be corrected prior to the registration audit. If the preliminary assessment audit is conducted early enough, it can serve the same purpose as the **assessment of current conformance** in Step 6. It also can provide a high-visibility jump start to get the preparation process under way. Whether it is done at Step 6 or at Step 13, it is an important part of the process. It uncovers any remaining areas of nonconformance so that they can be corrected before the registration audit. By identifying areas of nonconformance and taking the appropriate corrective action, an organization can approach the registration audit with a high degree of confidence and can minimize, if not eliminate, follow-up issues and their cost. The preliminary assessment audit can be a good investment. If an organization feels competent to go through Step 6 on its own, it should schedule the preliminary assessment at Step 13. Otherwise, the organization should select a registrar earlier and have the preliminary assessment audit done at Step 6.

As stated earlier, the preliminary assessment audit is optional. Many organizations and their registrars skip it if they are confident of conformance. However, all registrars will conduct a review of the organization's ISO 14000–related documentation at Step 13. The review is usually conducted at the registrar's facility, although site visits in connection with the documentation review are also not uncommon.

As a word of caution, you cannot count on the registrar providing a lot of help when a nonconformance is found, either through a preliminary assessment audit or through the documentation review. Remember that auditors must reveal any nonconformances but cannot tell *how to correct* them. ISO 9000 firms may find this to be even more closely observed for ISO 14000 than it is in ISO 9000 due to the potential for legal involvement. No matter what nonconformances are found by the registrar, the organization has the responsibility for determining how to correct them.

[Firms not seeking formal ISO 14000 registration should use internal auditors rather than the registrar in this step.]

## Step 14: Final Preaudit Touch-Up

The organization now has developed and implemented the EMS, and it has gained the information produced by the preliminary assessment audit (if conducted) and the registrar's documentation review. With time relatively short between this step and the formal registration audit, it can proceed with correcting the remaining deficiencies. Use the internal auditors to verify conformance as these remaining action items are completed. With the **final preaudit touch-up** completed, the registration audit should be uneventful. However, if one or more major nonconformances cannot be corrected in time, notify the registrar at once so that the registration audit can be rescheduled. It is pointless to go into an audit with a known major nonconformance because the registrar cannot grant registration until it is eliminated. It is better to postpone than to pay for an extra audit.

[Firms not seeking formal ISO 14000 registration should use internal auditors in place of the registrar in this step.]

## Step 15: Registration Audit

The registration audit is conducted according to the structured procedure described in chapter 6. If no major nonconformances are found and the auditors are satisfied that the implementation of the EMS is sound, the lead auditor will recommend that the registrar grant registration. After a review of the audit data and finding that all is in order, the registrar will award registration. The organization, even though registered, may still have to satisfy minor nonconformance reports and corrective action requests.

*Note*: If the organization has worked honestly and openly with the registrar and its own internal auditors, it would be unusual for a major nonconformance to be detected by the registration audit. Even without the optional preliminary assessment audit, the registrar's document review will usually reveal any major nonconformance. Both parties would be embarrassed to discover a major nonconformance at this point.

[Firms not seeking formal ISO 14000 registration should duplicate Step 15 using their internal auditors.]

## FOLLOW-UP TO REGISTRATION

After securing registration the organization must respond to any CARs that are still outstanding and advise the registrar of corresponding actions. The registrar may schedule a follow-up visit to verify the effectiveness of corrective actions or may choose to follow up at the next surveillance audit.

The registrar will close the CARs upon verification of satisfactory corrective action. Now that the organization is registered to ISO 14000, it is up to all employees, but especially management, to maintain registration. This requires careful and attentive monitoring of the EMS for continued suitability, adequacy, and effectiveness; for ways to improve it further (remember the commitment to continual improvement); and for scheduled internal audits to verify continued conformance with ISO 14001 and compliance with legal, regulatory, and other requirements as well as the effectiveness of the EMS. The registrar's auditors will return (usually at six-month intervals) to conduct their independent verification. At the three-year point, the registrar will conduct a full-scale reregistration audit.

## SUMMARY

1. The primary incentives that influence an organization's decision to implement ISO 14000 are as follows: customer demand, competitive advantage, desire to improve the EMS, public relations, influence on government, and need to streamline existing programs.

2. ISO 14004 gives the following reasons why an organization should implement an effective EMS: protect human health, and protect, maintain, and improve the environment. Another reason not given in ISO 14004 is that implementing an effective environmental management system is good business.

3. The benefits of third-party registration over self-declaration are as follows: credibility, reduced exposure to liability and litigation, accessibility to environmental insurance, fewer audits, and continuing attention by top management.

4. The most common reasons that firms fail the certification audit are as follows: failure to address all ISO 14001 requirements, failure to integrate continual improvement, failure to incorporate the five EMS principles, failure to determine and list environmental aspects, failure to implement targets and objectives, inadequate performance monitoring, inadequate internal auditing, failure to provide adequate document control, failure to maintain calibration of monitoring and measuring tools and equipment, failure to adhere to published procedures and practices, failure of actual employee credentials to match published job descriptions, lack of periodic reviews of the EMS, and breakdown of management responsibility.

5. Cost-minimization strategies include the following: minimizing the number of consultants, developing in-house consultants, developing the EMS in-house, developing documentation in-house, properly selecting a registrar, and controlling the timetable.

6. The fifteen steps recommended by the authors for seeking registration are as follows: secure commitment from the top, decide to proceed, form a steering committee, train the steering committee, select and train internal auditors, assess current conformance, plan registration preparation projects, select project teams, train project teams, activate project teams, collect feedback and monitor progress, select a registrar, perform preassessment audit and document review, conduct final preaudit touch-up, and complete registration audit.

7. After registration is awarded, outstanding CARs require a response. A follow-up visit may be scheduled to verify corrective action. Then there will be regularly scheduled internal audits to verify continued conformance. In addition the registrar's auditors will return periodically to conduct independent verification. A full-scale reregistration audit will be conducted at the three-year point.

## KEY CONCEPTS

| | |
|---|---|
| Assessment of current conformance | Continuing attention by top management |
| Commitment at the top | Credibility |
| Consultants | Decision to proceed |

Developing documentation in-house or out-of-house

Developing the EMS in-house or out-of-house

Environmental insurance

Fewer audits

Final preaudit touch-up

Internal auditor selection and training

Misinterpretation of ISO 14001 requirements

Overcontrolling through the EMS

Overdoing the development of ISO 14000 documentation

Plan preparation projects

Preliminary assessment audit and document review

Project feedback and monitoring

Project team activation

Project team selection

Reduced exposure to liability and litigation

Registrar selection

Registration audit

Registration timetable

Selecting a registrar

Self-declaration

Steering committee

Steering committee training

Team training

## REVIEW QUESTIONS

1. List the principal incentives that lead organizations to pursue ISO 14000 registration.
2. List at least five questions an organization can ask when trying to decide if it should seek ISO 14000 registration.
3. From the lessons learned with ISO 9000, we know that certain benefits will accrue from the implementation of ISO 14000. What are the benefits?
4. According to ISO 14004, why should an organization adopt the ISO 14000 EMS?
5. What is meant when it is said that an organization *self-declares* conformance to ISO 14000?
6. Explain the following benefits of pursuing registration through a registrar:
   - Credibility
   - Fewer audits
   - Continuing attention by top management
7. List ten common reasons why organizations fail to pass the ISO 14000 registration audit.
8. What is meant by "overdoing the development of ISO 14000 documentation"?
9. What is meant by "overcontrolling through the EMS"?
10. Explain three ways that organizations can minimize the *long-term* costs of ISO 14000 registration.
11. Develop a brief summary of each of the fifteen steps to ISO 14000 registration.
12. Determine if this statement is true: "ISO 14000 is a never-ending undertaking." Explain.

## CRITICAL-THINKING ACTIVITIES

The following activities may be assigned as individual, group, or discussion activities to be completed in class or out of class.

1. You have been asked to make a presentation to your company's senior management team to help its members decide the following question: Should our company implement ISO 14000?

2. Argue the case for self-declaration. Then, argue the case for registration. Which approach would you recommend to an organization considering ISO 14000 implementation?

3. Take all of the potential registration problems explained in this chapter and convert them into a checklist. Identify a company that has pursued ISO 14000 registration. Ask a representative of the company to indicate which problems on the checklist the company experienced. Compare your checklist with other students who completed this activity.

4. Assume that you are the environmental manager for a manufacturing company employing 1,500 people. The CEO has decided to pursue ISO 14000 registration but wants to minimize long-term costs. Develop a plan for minimizing the costs while simultaneously maximizing the benefits of ISO 14000 registration.

5. This activity should be undertaken by a team of students or the entire class, with different responsibilities assigned to different members and the overall project coordinated by a team leader. Create a fictitious company and use the fifteen-step model from this chapter to complete a hypothetical implementation. Record all activities, including hypothetical problems encountered and how they were solved. The finished product should be a handbook showing every aspect of the company's ISO 14000 implementation. Each team member should be able to document his or her knowledge of ISO 14000 for potential employers.

## DISCUSSION CASE

### The Company That Did It Wrong

A company known as ABC, Inc., decided that its position in the community would be enhanced by ISO 14000 registration. The company's CEO told his senior managers, "I'd like to be able to wave the ISO 14000 flag the next time some reporter comes snooping around asking questions about water pollution and fish kills." Then he told his vice president for engineering, "Select one of your best employees, and give him or her the responsibility for getting the company registered. I want to be registered *yesterday*, and I want to spend the least money possible. Also, find us an easy registrar."

ABC, Inc., failed to gain registration. As a result, the CEO fired his in-house ISO coordinator, whom he claimed had taken too much time and had wanted to spend too much money. The company is now involved in a multimillion-dollar lawsuit which alleges that ABC, Inc., knowingly and willfully polluted a public waterway by dumping untreated effluents.

## Discussion Questions

1. What went wrong with the ISO 14000 implementation at ABC, Inc.?

2. Analyze this case. What could and should have been done differently?

## ═══════ ENDNOTES ═══════

1. John Cascio, ed., *The ISO 14000 Handbook* (Fairfax, VA, and Milwaukee: CEEM Information Services and ASQ Press, 1996), p. 10.

2. Ibid.

3. ISO press release, ref.: 755, January 7, 1999.

4. ANSI/ISO 14004-1996, *Environmental Management Systems—General Guidelines on Principles, Systems and Supporting Techniques* (Milwaukee, 1996), Clause 0.2, p. viii.

5. ANSI/ISO 14001—1996, *Environmental Management Systems—Specification with Guidance for Use* (Milwaukee, 1996), Introduction, p. vii.

6. ANSI/ISO 14001—1996, Clause 4.3.2.

7. ISO press release, ref.: 756, February 1, 1999.

8. Peter R. Sholtes, *The Team Handbook* (Madison, WI: Joiner Associates, Inc., 1992).

9. David Goetsch and Stanley Davis, *Quality Management: Introduction to Total Quality Management for Production, Processing, and Services*, 3d ed. (Upper Saddle River, NJ: Prentice-Hall, 2000), particularly Chapter 10.

# Other ISO Standards, Issues, and Developments

## ISO 9000

ISO 14000 fundamentally requires an environmental management system—a system that enables organizations to reliably and consistently meet legal, regulatory, and other environmental requirements and improve environmental performance over time. ISO 14000 does not impose performance requirements. Rather, it provides a management system through which the organization can routinely meet all performance requirements with which the firm must comply. ISO 9000 is similar. Although it is an International Quality Standard, ISO 9000 does not establish levels of quality performance. ISO 9000 does require a *quality management system* and provides an effective model for it. The objective of ISO 9000 is consistent quality performance, at a level established by the organization. Notice that, unlike ISO 14000 where government entities usually establish environmental performance standards, ISO 9000 deals with an unregulated concept—quality. Under ISO 9000 each organization decides for itself the level of quality performance that will satisfy its customers. Then, through the **ISO 9000 quality management system**, the organization delivers that level of quality time after time after time. ISO 9000 is a family of standards that represent a worldwide consensus (as determined by Technical Committee 176) concerning the best quality management practices. The aim of ISO 9000 is to enable organizations to consistently meet their customers' quality requirements.

ISO 9000 is applicable to all organizations that produce goods or provide services, large or small, public or private. To cover such a broad range of organizations, ISO purposely established the requirements for developing and operating a quality management system but left to each organization the question of *how* the requirements are to be met. In other words, ISO 9000 tells the organization what it must do in terms of a quality management system, and the organization determines how it will do it. This is the same philosophy used in ISO 14000, which is only natural since ISO 9000 had a ten-year head start, and the lessons learned in ISO 9000 have been implemented in ISO 14000.

The current ISO 9000 family is comprised of eleven documents as listed below:

### Guidelines for Quality Management and Quality Assurance Standards

ISO 9000-1:1994    Quality Management and Quality Assurance Standards—Guidelines for Selection and Use

ISO 9000-2:1993    Quality Management and Quality Assurance Standards—Part 2: Generic Guidelines for the Application of **ISO 9001**, **ISO 9002**, and **ISO 9003**

ISO 9000-3:1991    Quality Management and Quality Assurance Standards—Part 3: Guidelines for the Application of ISO 9001 to the Development, Supply, and Maintenance of Software

ISO 9000-4:1993    Quality Management and Quality Assurance Standards—Part 4: Guide to Dependability Program Management

### Standards for Quality Systems and Models for Quality Assurance

ISO 9001:1994    Quality Systems—Model for Quality Assurance in Design, Development, Production, Installation, and Servicing

ISO 9002:1994    Quality Systems—Model for Quality Assurance in Production, Installation, and Servicing

ISO 9003:1994    Quality Systems—Model for Quality Assurance in Final Inspection and Test

### Guidelines for Quality Management and Quality System Elements

ISO 9004-1:1994    Quality Management and Quality System Elements—Guidelines

ISO 9004-2:1991    Quality Management and Quality System Elements—Part 2: Guidelines for Services

ISO 9004-3:1993    Quality Management and Quality System Elements—Part 3: Guidelines for Processed Materials

ISO 9004-4:1993    Quality Management and Quality System Elements—Part 4: Guidelines for Quality Improvement

This family of standards provides both the requirements and guidance for an effective quality management system and its elements. (ISO will issue an updated version of the standard in late 2000. It will be designated ISO 9000:2000.)

Essentially everything that has been said in this book about registration and auditing to the ISO 14000 Standard also applies to ISO 9000. In concept they are very similar. ISO 14000 is aimed at environmental performance, driven by legal and regulatory agencies; likewise, the ISO 9000 Standard is concerned with quality performance, driven by customers.

For more information on ISO 9000, review the companion book to this text, *Understanding and Implementing ISO 9000 and ISO Standards*, 1998, Prentice-Hall.

## OCCUPATIONAL HEALTH AND SAFETY

Three fundamental issues affect all organizations that must compete in the global market:

- quality of products and services
- protection of the environment
- welfare of employees

ISO has addressed two of the three issues with ISO 9000 and ISO 14000. It seems appropriate, and in fact it has been anticipated, that an International Standard for occupational safety and health be developed under ISO and integrated with ISO 9000 and ISO 14000. The designation ISO 18000 has been reserved for a safety and health standard. However, to date, no technical committee has been established for that purpose.

In the United States the Occupational Safety and Health Administration (**OSHA**) has the statutory authority to develop, proclaim, and enforce occupational safety and health standards. Many other countries have similar organizations and some do not. The result is that standards for occupational health and safety exist to protect employees in many nations, yet in many others the issue is ignored. An International Standard that is based on the management system models provided by ISO 9000 and ISO 14000 and that yields the best practices for occupational safety and health universally would be beneficial to organizations, employees, and consumers around the world.

Since 1991 OSHA has had a Memorandum of Understanding (MOU) concerning cooperation with the American National Standards Institute (**ANSI**).[1] Through this MOU, ANSI is to assist and support OSHA in the development of national consensus standards concerning occupational safety and health. ANSI is also to provide OSHA with proposed drafts of international safety and health standards proclaimed by ISO and other agencies, for the purpose of receiving OSHA comments as part of the overall position of the United States on the standards. Under the MOU, OSHA is to participate on ANSI-accredited standards committees.

What does this mean? Simply that if ISO were to establish a technical committee to draft an International Standard for occupational safety and health, the mechanism is already in place for the United States to participate. ANSI, as the United States member of ISO, with OSHA as its subject matter expert, would certainly become part of an ISO technical committee charged with developing an International Standard for occupational safety and health. Occupational safety is discussed in depth in *Occupational Safety and Health*, 3d Edition, 1999, by David L. Goetsch.

## ANTICIPATED ISSUES AND DEVELOPMENTS

### Merging Management System Standards

It has been anticipated that eventually there will be three international management system standards—quality, environment, and occupational health and safety—and that they will be integrated into a single standard. As we have seen, two of the three International Standards, ISO 9000 and ISO 14000, currently exist and have been accepted worldwide. The third, ISO 18000, remains to be developed. Nevertheless, the concept of **merging management system standards** is active. The advantages of having the two standards become one are considerable. The possibilities include a single registration taking advantage of common and overlapping procedures and organizational structure. In addition, a single audit could cover both quality and environmental management systems and eliminate duplication in procedures, structure, and methods, and lower overall costs of registration.

On the other hand, merging the two standards would pose a problem to the organization that is prepared for registration to one standard but not the other. In this case, registration would be delayed until the organization could satisfy both. For some organizations, this could impose an unacceptable delay or added cost. This may prevent ISO from merging the standards completely. However, many of the advantages of merging may be achieved without a complete merger.

This appears to be the path that ISO is following. A Technical Advisory Group (TAG 12), formed to look into merging ISO 9000 and ISO 14000, recommended that the standards not be merged, but be made more compatible. Specific recommendations for the standards include:[2]

- Relevant terms and definitions should be identical and there should be consistent use of terminology in both families of standards.
- Management system standards in the two families should be compatible and, as far as possible, aligned.
- Auditing standards in the two families should be integrated as far as possible to consist of a common core document with accompanying separate modules on quality and environment.

ISO's Technical Management Board (TMB) has tasked the technical committees responsible for ISO 9000 (TC 176) and ISO 14000 (TC 207) to implement TAG 12's recommendations. High priority is given to the development of a joint auditing standard.[3] This would enable registrars to use a single surveillance audit to cover the requirements of both standards, and it conceivably could even include the registration audit. This would cut the number of audits of the organization by half and would be less expensive.

TC 207 and TC 176 have established a Joint Coordination Group to achieve compatibility between the two standards.[4] **Compatibility** is defined by ISO as the "suitability of products, processes or services for use together under specific conditions to fulfill relevant requirements without causing unacceptable interactions." This means that the two standards must be capable of "living together harmoniously."

For any of this to be possible, the periodic revisions for the two standards must occur simultaneously, or one will always be out of step with the other, thereby making joint auditing difficult and possibly adding work for the organization. ISO has told the two technical committees to synchronize their revisions.[5]

Finally, ISO has set up a committee to develop guidelines for a common methodology for the drafting of management system standards to facilitate compatibility, lower costs, and other benefits.[6]

In summary, the merging of ISO 9000 and ISO 14000 will not occur anytime soon, if ever. Nor is an ISO International Standard for occupational safety and health likely in the near future. However, revisions to both ISO 14000 and ISO 9000 will make them more compatible and more closely aligned in order to facilitate implementation and common auditing.

## Integrity of Management System Registrations

The authors have long held the view that the current system of accrediting registrars and certifying organizations to ISO 9000 and ISO 14000 is flawed. The integrity of the certification itself and the value of the standards is at stake. In other chapters of this book organizations have been cautioned to employ registrars with integrity, rather than registrars that might "look the other way." It is not known how widespread this problem is, but it does exist. It will continue to exist until registrars themselves are subjected to audits of their performance, something that is not done now. ISO 9000 and ISO 14000 are too important to let a few self-serving registrars render them valueless by damaging the integrity of the certification system. For more information on the **integrity of management system registrations**, refer to the ISO 9000 companion book to this text.[7]

## SUMMARY

1. ISO 14000 is an International Standard requiring an environmental management system that enables organizations to meet legal, regulatory, and other environmental requirements while simultaneously improving performance over time.

2. ISO 9000 also requires an effective management system (for quality) and is an International Standard. While ISO 14000 mandates adherence to both legal/regulatory and self-imposed performance requirements, ISO 9000 requires adherence only to self-imposed performance requirements. There are no mandatory national or international quality-level standards except those imposed by a competitive marketplace.

3. There is ongoing speculation about the development of an International Standard for employee health and safety. Thus far the idea is merely speculation, and it is likely to remain speculation for the foreseeable future. Should the speculation ever become reality, the United States is well positioned to play a key role in the development of the standard. The American National Standards Institute (ANSI) represents the United States in ISO. ANSI has a Memorandum of Understanding (MOU) with the Occupational Safety and Health Administration (OSHA), the organization with statutory authority in the United States for developing and enforcing standards for

occupational safety and health. Through the MOU, ANSI is to assist OSHA in developing a national consensus concerning occupational safety and health standards. In addition, OSHA is to participate on ANSI-accredited standards committees.

4. Although the complete merger of ISO's quality and environmental standards is not likely to occur in the foreseeable future, a merger of management systems might happen. In order for this to happen, the following action is necessary:

- Terms and definitions will have to be made consistent.
- Management system standards will have to be made compatible.
- Auditing standards will have to be integrated as far as possible.

## KEY CONCEPTS

ANSI

Compatibility

Integrity of management system registration

ISO 9000 quality management system

ISO 9001

ISO 9002

ISO 9003

Merging management system standards

OSHA

## REVIEW QUESTIONS

1. What is ISO 9000?
2. Distinguish between ISO 9000 and ISO 14000.
3. Which standard from the ISO 9000 "family" would an organization use for assuring quality in final inspection and test?
4. What are the three fundamental issues which affect all organizations that must compete in the global marketplace?
5. What organization in the United States has statutory authority to develop, proclaim, and enforce occupational safety and health standards?
6. What three actions will be necessary in order to make the ISO 9000 and ISO 14000 standards more compatible?
7. Explain the "integrity" issue as it relates to ISO registration.

## CRITICAL-THINKING PROBLEMS

The following activities may be assigned as individual, group, or discussion activities to be completed in class or out of class.

1. Identify as many advantages as possible for an organization that has already completed ISO 9000 registration and wants to pursue ISO 14000 registration.

2. The most widely applicable OSHA standards are the General Industry Standards. They are found in 29 C.F.R. 1910 (read as *Title 29, Code of Federal Regulations, Part 1910*). Locate a copy of OSHA's General Industry Standards and answer the following questions:

   • Based on the contents of 29 C.F.R. 1910, what problems would you anticipate in trying to merge OSHA's standards with ISO 14000?

   • What international inhibitors would you anticipate (i.e., different attitudes toward air quality, machine use, sanitation, etc.)?

3. Assume that you have been asked to attend a meeting of your company's senior management team. Your assignment is to convince these executives to hire the toughest ISO 14000 registrar that can be found instead of one with a reputation for "looking the other way." Validate the case for hiring the registrar with integrity.

## DISCUSSION CASE

Magna Technologies, Inc., is an ISO 9000–registered company. It also has an excellent record in the area of occupational safety and health, or at least it used to. Magna's CEO planned to pursue ISO 14000 registration now that ISO 9000 registration had been achieved. In his mind, ISO 9000, occupational safety, and ISO 14000 all fit neatly under one umbrella. Under the rationale of "rightsizing," the CEO promoted his quality manager to Director of Quality, Safety, and Environmental Management.

At the same time, the CEO eliminated Magna's safety manager and environmental manager. In the CEO's opinion, he was promoting a competent professional who had done an excellent job of getting the company registered to ISO 9000. He told the other senior managers that " . . . ISO 9000, ISO 14000, and occupational safety are all about compliance. Surely one person can handle all three types of compliance."

Before long the company's safety record began to slip. To complicate matters, the ISO 14000 preparation did not fall in place the way ISO 9000 preparation had. In fact, ISO 14000 registration had to be postponed indefinitely while Magna personnel dealt with quality and safety problems that had begun to plague the company.

### Discussion Questions

1. What went wrong at Magna Technologies, Inc.?

2. Is it reasonable to expect one person to have expertise in quality, safety, and environmental management?

3. If you were asked to advise Magna's CEO, what would you recommend?

## ENDNOTES

1. OSHA Memorandums of Understanding, *Cooperation Between OSHA and ANSI*, May 21, 1991.

2.  ISO press release, *Advisory Group Recommends Actions for Greater Compatibility, but No Merging of ISO 9000 and ISO 14000,* undated (1998).

3.  ISO press release, *ISO Places "Rush Order" for Joint ISO 9000 and ISO 14000 Audit Standard,* September 16, 1998.

4.  Ibid.

5.  Ibid.

6.  Ibid.

7.  Goetsch and Davis, *Understanding and Implementing ISO 9000 and ISO Standards* (Upper Saddle River, NJ: Prentice-Hall, 1998), p. 187.

# CHAPTER TEN

# ISO 14000 Checklist

- Purpose of the Checklist
- Checklist Format
- How To Use the Checklist
- ISO 14000 Checklist

The format of chapter 10 differs from that of previous chapters. Chapters 1–9 concluded with teaching and learning aids such as review questions, critical-thinking activities, and case studies. Chapter 10 as a whole is a teaching and learning activity. It contains the ISO 14000 Standard in checklist form. Focus your attention on the standard and learning how to use it.

## PURPOSE OF THE CHECKLIST

The ISO 14000 Checklist is intended to assist organizations in self-assessing their readiness for registration. It may also be used by organizations intending to self-declare conformance to ISO 14000. The checklist is not a shortcut to conformance. The organization must develop and implement its conforming environmental management system according to ISO 14001, upon which the checklist is based. The checklist will, however, assist organizations in keeping track of EMS development and implementation by providing a convenient means of tracking progress. It also will reveal more clearly than the standard itself what remains to be done.

## CHECKLIST FORMAT

The checklist corresponds directly with the requirements of the ISO 14001 Standard.[1] It is organized by the same headings and subheadings as the standard, and all requirements in the standard are listed.

## HOW TO USE THE CHECKLIST

The checklist can be used as an internal tool for preparing an organization for ISO 14000 registration or for self-declaration of ISO 14000 conformance. For each question an affirmative answer indicates, but does not prove, that the organization has satisfied the requirement in question.

While using the checklist, refer to other chapters for clarification, especially chapters 3, 4, and 5. ISO 14004 and Annex A to ISO 14001 will also be useful for this purpose.

## ISO 14000 CHECKLIST

*Note*: As stated earlier, each checklist item relates directly to the specified clause of ISO 14001, Section 4, Environmental Management System Requirements.

### 4 Environmental Management System Requirements

#### 4.1 General Requirements

_____ 1. Has the organization established an environmental management system to meet all requirements of Clause 4?

_____ 2. Is the environmental management system maintained; i.e., is it in daily use, is it routinely followed, and are its requirements enforced by management?

#### 4.2 Environmental Policy

_____ 3. Has top management defined the organization's environmental policy?

_____ 4. Is the environmental policy appropriate to the nature of the organization's activities, products, or services?

_____ 5. Is the policy appropriate to the scale of the organization's activities, products, or services?

_____ 6. Is the policy appropriate to the actual or possible environmental impacts of the organization's activities, products, or services?

_____ 7. Does the environmental policy include a management commitment to the prevention of pollution?

_____ 8. Does the policy include a management commitment to continual improvement?

_____ 9. Does the policy include a management commitment to comply with all relevant environmental laws and regulations?

_____ 10. Does the policy include a management commitment to comply with other environmental requirements to which the organization subscribes?

_____ 11. Does the environmental policy provide a workable framework for setting environmental objectives and targets?

_____ 12. Does the policy provide a workable framework for reviewing environmental objectives and targets?

____ 13. Is the environmental policy documented?

____ 14. Has the policy been implemented and maintained (i.e., put into daily use)?

____ 15. Has the policy been communicated to *all* employees?

____ 16. Is the environmental policy available to the public?

## 4.3 Planning

### 4.3.1 Environmental Aspects

____ 17. Has the organization established (a) procedure(s) to identify the environmental aspects of its activities, products, or services that it can control and over which it can be expected to have an influence?

____ 18. Does the organization maintain, i.e., routinely use, this (these) procedure(s)?

____ 19. Has the organization determined through the use of this (these) procedure(s) which of its environmental aspects have, or can have, significant impacts on the environment?

____ 20. Does (Do) the organization's procedure(s) ensure that the aspects related to significant environmental impacts are always considered in setting environmental objectives?

____ 21. Does the organization routinely keep information relative to its environmental aspects up-to-date?

### 4.3.2 Legal and Other Requirements

____ 22. Does the organization have a procedure to identify legal and regulatory requirements applicable to the environmental aspects of its activities, products, or services?

____ 23. Does the organization have a procedure to identify other environmental requirements to which it subscribes?

____ 24. Does the organization have a procedure that assures access to all appropriate legal, regulatory, and other applicable environmental requirements?

____ 25. Does the organization routinely use (maintain) this (these) procedure(s)?

### 4.3.3 Objectives and Targets

____ 26. Does each relevant function and level of the organization have current, documented environmental objectives and targets?

____ 27. Are the relevant functions and levels of the organization routinely working to achieve the current, documented environmental objectives and targets?

____ 28. In establishing and reviewing its environmental objectives and targets, does the organization consider all of the following?

____    a. legal, regulatory and other environmental requirements

____    b. its significant environmental aspects

____    c. its technological options

____    d. its financial, operational and business requirements

____    e. the views of interested parties, both in and out of the organization

____ 29. Are the organization's objectives and targets consistent with its environmental policy?

____ 30. Are the organization's objectives and targets consistent with its commitment to prevent environmental pollution?

### 4.3.4 Environmental Management Program(s)

____ 31. Has the organization established (a) program(s) for achieving its environmental objectives and targets?

____ 32. Within this (these) program(s), is there a designation of responsibility for achieving objectives and targets at each relevant function and level of the organization?

____ 33. Within this (these) program(s), is the means by which the objectives and targets are to be achieved stated and explained?

____ 34. Within this (these) program(s), is a schedule provided for achieving the objectives and targets?

____ 35. Is (are) this (these) program(s) routinely used and followed for achieving the organization's objectives and targets?

____ 36. To ensure that environmental management applies to new developments, new or modified activities, new or modified products, and new or modified services, does the organization do one or both of the following?

____    a. Develop new environmental management programs.

____    b. Amend relevant existing programs.

## 4.4 Implementation and Operation

### 4.4.1 Structure and Responsibility

____ 37. Has the organization defined the roles, responsibilities, and authorities of its employees in connection with environmental management?

____ 38. Have these roles, responsibilities, and authorities been documented through organizational charts or other means?

____ 39. Have these roles, responsibilities, and authorities been communicated to the organization's employees?

____ 40. Has management provided the resources essential to the implementation and control of the EMS, including the following?

____    a. human resources (people, manpower)

____    b. required specialized skills (from within or from outside)

____    c. essential technology (equipment, processes, methods)

_____    d. financial resources

_____ 41. Has top management appointed one or more specific *Management Represen-tatives*?

_____ 42. In addition to any other responsibilities already held, has management defined the environmental role(s) of the Management Representative(s) to include the responsibility and authority for:

_____    a. ensuring that ISO 14000 EMS requirements are established?

_____    b. ensuring that ISO 14000 EMS requirements are implemented?

_____    c. ensuring that the conforming EMS is properly maintained and used?

_____    d. reporting on the performance of the EMS to top management for the purposes of review and as a basis for EMS improvement?

### 4.4.2 Training, Awareness and Competence

_____ 43. Does the organization require that all employees whose work may create a significant impact on the environment receive appropriate training?

_____ 44. Has the organization identified training needs?

_____ 45. Does the organization have procedures to make employees at each relevant function and level aware of:

_____    a. the importance of conformance with the organization's environmental policy and procedures and with the requirements of the EMS?

_____    b. the actual or potential significant environmental impacts of their work activities?

_____    c. the environmental benefits of improved personal performance?

_____    d. their roles and responsibilities in achieving conformance with the organization's environmental policy and procedures and with the requirements of the EMS?

_____    e. their roles and responsibilities concerning the requirements for emergency preparedness and response?

_____    f. the potential consequences of departure from specified operating procedures?

_____ 46. Are these awareness procedures routinely used?

_____ 47. Does the organization ensure that employees performing tasks which can cause significant environmental impacts be competent to perform them on the basis of education, training, and/or experience?

### 4.4.3 Communication

_____ 48. Does the organization have communication procedures for dealing internally with concerns and questions about the organization's environmental aspects and the EMS?

_____ 49. Do these procedures support communication between the various functions and levels of the organization?

_____ 50. Does the organization have communication procedures for receiving, documenting, and responding to relevant environmental communications from interested parties outside the organization?

_____ 51. Has the organization considered processes for external communication on its significant environmental aspects and whether it should implement such processes?

_____ 52. Has the organization documented these considerations and decisions?

### 4.4.4 Environmental Management System Documentation

_____ 53. Has the organization documented information (in paper or electronic form) that describes the core elements of the EMS and how they interact with each other?

_____ 54. Has the organization documented information (in paper or electronic form) that provides direction to EMS-related documentation, and has the organization ensured that when employees need documentation they can locate it?

_____ 55. Is this information maintained up-to-date, and is it routinely used by employees?

### 4.4.5 Document Control

_____ 56. Has the organization established procedures for controlling all documents required by ISO 14001?

_____ 57. Do these procedures ensure that all documents required by ISO 14001 can be located?

_____ 58. Do these procedures ensure that all documents required by ISO 14001 are periodically reviewed, revised as necessary, and approved for adequacy by authorized personnel?

_____ 59. Do these procedures ensure that the *current versions* of all documents required by ISO 14001 are available at all critical locations? ("Critical locations" means where work operations essential to the effective functioning of the EMS take place.)

_____ 60. Do these procedures ensure that *obsolete documents* are promptly removed from all points of issue and points of use, or do the procedures ensure in some other manner that *obsolete documents* cannot be used unintentionally?

_____ 61. Do these procedures ensure that any obsolete documents retained for any reason (e.g., legal or knowledge preservation) are suitably and clearly identified as obsolete?

_____ 62. Has the organization established procedures and responsibilities for creating and modifying the various types of documentation?

_____ 63. Do the procedures for creating and modifying documentation require that documents:

_____     a. be legible?

_____     b. carry the origination date?

_____     c.  carry the date of any revision?

_____     d.  be readily identifiable?

_____     e.  be approved by signature of authorized individual(s)?

_____  64. Are all of these document-control procedures routinely used and rigorously enforced by the organization?

_____  65. Does the organization maintain (file, store, protect, disperse for use, retrieve, etc.) its controlled documentation in an orderly manner?

_____  66. Does the organization retain its controlled documentation for a specified period (determined by legal or regulatory requirements or by the organization itself)?

### 4.4.6 Operational Control

_____  67. To ensure that its environmental policy and its targets and objectives can be met, has the organization identified its operations and activities that are associated with its significant environmental aspects?

_____  68. Has the organization planned these operations and activities, including maintenance, to ensure that they are carried out under specified (controlled) conditions by:

_____     a.  establishing documented procedures to cover situations where the absence of documented procedures could lead to deviations from the environmental policy and the targets and objectives?

_____     b.  stipulating operating criteria (step-by-step work instructions, parametric readings, etc.) in the procedures?

_____     c.  establishing procedures related to identifiable significant environmental aspects of goods and services used by the organization?

_____     d.  communicating relevant procedures and requirements to the suppliers and contractors who provide these goods and services?

_____  69. Are all of these operational control procedures routinely used and rigorously enforced by the organization?

### 4.4.7 Emergency Preparedness and Response

_____  70. Has the organization established emergency preparedness and response procedures to identify potential for environmentally related accidents and emergency situations, and has it developed procedures to respond to such accidents and emergency situations?

_____  71. Do these procedures address the prevention and/or mitigation of environmental impacts that may result from such accidents?

_____  72. Does the organization routinely use and rigorously enforce its emergency preparedness and response procedures?

_____ 73. Does the organization review (and revise when necessary) its emergency preparedness and response procedures (especially after the occurrence of accidents or emergency situations)?

_____ 74. Does the organization periodically test these procedures where practicable?

## 4.5 Checking and Corrective Action

### 4.5.1 Monitoring and Measurement

_____ 75. Does the organization have documented procedures to monitor and measure the key characteristics of its operations and activities that can have a significant impact on the environment?

_____ 76. Do these documented procedures require that monitoring and measuring of its key characteristics be done on a regular basis?

_____ 77. Do these procedures require the recording of information to track:

_____     a. environmental performance?

_____     b. relevant operational controls?

_____     c. conformance with the organization's environmental objectives and targets?

_____ 78. Are these documented procedures routinely followed?

_____ 79. Does the organization calibrate and maintain its monitoring equipment according to an established process and schedule?

_____ 80. Does the organization have a procedure requiring the retention of records of monitoring equipment maintenance and calibration activity?

_____ 81. Are these records being retained as required by the procedure?

_____ 82. Does the organization have a documented procedure for periodically evaluating compliance with relevant environmental legislation and regulations?

_____ 83. Is this procedure routinely followed?

### 4.5.2 Nonconformance and Corrective and Preventive Action

_____ 84. Has the organization established procedures for defining responsibility and authority for:

_____     a. handling and investigating nonconformance?

_____     b. taking action to mitigate any impacts caused by nonconformances?

_____     c. initiating and completing corrective action?

_____     d. initiating and completing preventive action?

_____ 85. Does the organization ensure that any corrective or preventive action taken to eliminate the causes of actual or potential nonconformances are:

_____ a. appropriate to the magnitude of the problems and

_____ b. commensurate with the environmental impact encountered (or which is possible)?

_____ 86. When corrective and/or preventive actions involve documented procedures, does the organization implement and record the necessary changes to those procedures (i.e., are the procedures formally modified)?

### 4.5.3 Records

_____ 87. Has the organization established procedures for environmental records, including training records and the results of audits and reviews, that covers their:

_____ a. identification?

_____ b. maintenance (i.e., collection, indexing, filing, storage, retrieval, retention, control, safekeeping)?

_____ c. disposition?

_____ 88. Do these procedures require that:

_____ a. environmental records be legible?

_____ b. environmental records be readily identifiable?

_____ c. environmental records be traceable to the activity, product, or service involved?

_____ d. environmental records be protected against damage, deterioration, or loss?

_____ e. environmental records be readily retrievable?

_____ f. retention times (periods) of environmental records be established and recorded?

_____ 89. Are these procedures routinely followed and enforced?

_____ 90. Does the organization maintain records under these procedures to demonstrate conformance to the requirements of ISO 14000, including, as appropriate:

_____ a. legislative and regulatory requirements?

_____ b. relevant permits?

_____ c. environmental aspects and their associated impacts?

_____ d. product information?

_____ e. process information?

_____ f. environmental training activity?

_____ g. inspection, calibration, and maintenance activity?

_____ h. monitoring data?

_____ i. details of nonconformance (incidents, complaints, and follow-up action)?

_____ j.  environmental audits (internal and external)?

_____ k.  management reviews?

_____ l.  supplier and contractor information?

_____ m. emergency preparedness and response information?

### 4.5.4 Environmental Management System Audit

_____ 91. Has the organization established (a) program(s) and procedures for periodic internal EMS audits?

_____ 92. Are these programs and procedures designed to:

_____ a.  determine whether or not the EMS conforms to the organization's planned arrangements for the EMS, and ISO 14000 requirements?

_____ b.  determine whether the EMS has been properly implemented and maintained?

_____ c.  provide audit result information to management?

_____ 93. Do the organization's internal audit programs and schedules of audits take into account the environmental importance (significance) of the activity concerned?

_____ 94. Do the organization's internal audit programs and schedules of audits take into account the results of previous audits (internal or external)?

_____ 95. Do the organization's internal EMS audit procedures cover:

_____ a.  designating the organizational activities and areas to be considered in audits?

_____ b.  establishing the frequency of audits?

_____ c.  establishing audit methodology?

_____ d.  assigning responsibilities associated with managing and conducting audits (recognizing the need that persons selected as auditors must be in a position to conduct audits impartially and objectively)?

_____ e.  ensuring auditor competence?

_____ f.  communicating audit results?

### 4.6 Management Review

_____ 96. Does the organization's top management periodically review the EMS to ensure its continuing suitability, adequacy, and effectiveness?

_____ 97. Are such reviews carried out according to a schedule that top management has determined (as opposed to ad hoc reviews)?

_____ 98. Does the organization's management review process ensure that the information necessary for management's evaluation (for suitability, adequacy, and effectiveness) is collected and presented?

_____ 99. Are the management reviews documented?

_____ 100. In light of EMS audit results, the extent to which objectives and targets have been met, concerns among relevant interested parties, changing circumstances, and the commitment to continual improvement, do the organization's management reviews address the possible need for changes to the organization's:

_____  a. environmental (or other) policy?

_____  b. environmental objectives and /or targets?

_____  c. other elements of the EMS (procedures, processes, etc.)?

## ENDNOTES

1. ANSI/ISO 14001-1996, Section 4, pp. 2–5.

# ISO Member Body Roll (Full Members)

The following membership list was provided by the International Organization for Standardization as of September 16, 1998. The entries are alphabetical by nation (in bold print). Each entry includes the name of the national standards organization (abbreviation in parenthesis) and the organization's address.

**Albania**
Drejtoria e Standardizimit dhe Cilesise (DSC)
Rruga Mine Peza Nr. 143/3, Tirana

**Algeria**
Institut Algérien de Normalisation (IANOR)
5, rue Abou Hamou Moussa, B.P. 403 - Centre de tri, Alger

**Argentina**
Instituto Argentino de Normalización (IRAM)
Chile 1192, 1098 Buenos Aires

**Armenia**
Department for Standardization, Metrology and Certification (SARM)
Komitas Avenue 42/2, 375051 Yerevan

**Australia**
Standards Australia (SAA)
1 The Crescent, Homebush - N.S.W. 2140

**Austria**
Österreichisches Normungsinstitut (ON)
Heinestrasse 38, Postfach 130, A-1021 Wien

**Bangladesh**
Bangladesh Standards and Testing Institution (BSTI)
116/A, Tejgaon Industrial Area, Dhaka-1208

### Belarus
Committee for Standardization, Metrology and Certification (BELST)
Starovilensky Trakt 93, Minsk 220053

### Belgium
Institut belge de normalisation (IBN)
Av. De la Brabanconne 29, B-1000 Bruxelles

### Bosnia and Herzegovina
Institute for Standardization, Metrology and Patents (BASMP)
Hamdije Cemerlica 2, (ENERGOINVEST building), CH-71000 Sarajevo

### Brazil
Associacao Brasileira de Normas Técnicas (ABNT)
Av. 13 de Maio, no. 13, 28o andar, 20003-900 - Rio de Janeiro-RJ

### Bulgaria
Committee for Standardization and Metrology (BDS)
21, 6th September Str., 1000 Sofia

### Canada
Standards Council of Canada (SCC)
45 O'Connor Street, Suite 1200, Ottawa, Ontario K1P 6N7

### Chile
Instituto Nacional de Normalización (INN)
Matías Cousiño 64 - 60 piso, Casilla 995 - Correo Central, Santiago

### China
China State Bureau of Technical Supervision (CSBTS)
4, Zhichun Road, Haidian District, P.O. Box 8010, Beijing 100088

### Colombia
Instituto Colombiano de Normas Técnicas y Certificación (ICONTEC)
Carrera 37 52-95, Edificio ICONTEC, P.O. Box 14237, Santafé de Bogotá

### Costa Rica
Instituto de Normas Técnicas de Costa Rica (INTECO)
P.O. Box 6189-1000, San José

### Croatia
State Office for Standardization and Metrology (DZNM)
Ulica grada Vukovara 78, 10000 Zagreb

### Cuba
Oficina Nacional de Normalización (NC)
Calle E No. 261 entre 11 y 13, Vedado, La Habana 10400

### Cyprus
Cyprus Organization for Standards and Control of Quality (CYS)
Ministry of Commerce, Industry and Tourism, Nicosia 1421

**Czech Republic**
> Czech Standards Institute (CSNI)
> Biskupsky dvur 5, 110 02 Praha 1

**Denmark**
> Dansk Standard (DS)
> Kollegievej 6, DK-2920 Charlottenlund

**Ecuador**
> Instituto Ecuatoriano de Normalización (INEN)
> P.O. Box 17-01-3999, Quito

**Egypt**
> Egyptian Organization for Standardization and Quality Control (EOS)
> 2 Latin America Street, Garden City, Cairo

**Ethiopia**
> Ethiopian Authority for Standardization (EAS)
> P.O. Box 2310, Addis Ababa

**Finland**
> Finnish Standards Association (SFS)
> P.O. Box 116, FIN-00241 Helsinki

**France**
> Association francaise de normalisation (AFNOR)
> Tour Europe, F-92049 Paris La Défense Cedex

**Germany**
> Deutsches Institut fûr Normung (DIN)
> Burggrafenstrasse 6, D-10772 Berlin

**Ghana**
> Ghana Standards Board (GSB)
> P.O. Box M 245, Accra

**Greece**
> Hellenic Organization for Standardization (ELOT)
> 313, Acharnon Street, GR-111 45 Athens

**Hungary**
> Magyar Szabványûgyi Testûlet (MSZT)
> Üllöi út 25, Pf. 24., H-1450 Budapest 9

**Iceland**
> Icelandic Council for Standardization (STRI)
> Keldnaholt, IS-112 Reykjavik

**India**
> Bureau of Indian Standards (BIS)
> Manak Bhavan, 9 Bahadur Shah Zafar Marg, New Delhi 110002

### Indonesia
Badan Standardisasi Nasional (BSN)
c/o Pusat Standardisasi - LIPI, Jalan Jend. Gatot Subroto 10, Jakarta 12710

### Iran
Institute of Standards and Industrial Research of Iran (ISIRI)
P.O. Box 31585-163, Karaj

### Ireland
National Standards Authority of Ireland (NSAI)
Glasnevin, Dublin-9

### Israel
Standards Institution of Israel (SII)
42 Chaim Levanon Street, Tel Aviv 69977

### Italy
Ente Nazionale Italiano di Unificazione (UNI)
Via Battistotti Sassi 11/b, I-20133 Milano

### Jamaica
Jamaica Bureau of Standards (JBS)
6 Winchester Road, P.O. Box 113, Kingston 10

### Japan
Japanese Industrial Standards Committee (JISC)
c/o Standards Department, Ministry of International Trade and Industry
1-3-1, Kasumigaseki, Chiyoda-ku, Tokyo 100

### Kenya
Kenya Bureau of Standards (KEBS)
P.O. Box 54974, Nairobi

### Korea, Democratic People's Republic of
Committee for Standardization of the Democratic People's Republic of Korea (CSK)
Zung Gu Yok Seungli-Street, Pyongyang

### Korea, Republic of
Korean National Institute of Technology and Quality (KNITQ)
2, Joongang-dong, Kwachon, Kyunggi-do 427-010

### Libyan Arab Jamahiriya
Libyan National Centre for Standardization and Metrology (LNCSM)
Industrial Research Centre Building, P.O. Box 5178, Tripoli

### Luxembourg
Service de l'Energie de l'Etat, Department Normalisation (SEE)
34 avenue de la Porte-Neuve, B.P. 10, L-2010 Luxembourg

**Malaysia**
> Department of Standards Malaysia (DSM)
> 21st Floor, Wisma MPSA, Persiaran Perbandaran, 40675 Shah Alam
> Selangor Darul Ehsan

**Mauritius**
> Mauritius Standards Bureau (MSB)
> Moka

**Mexico**
> Dirección General do Normas (DGN)
> Calle Puente de Tecamachalco No 6, Lomas de Tecamachalco
> Sección Fuentes, Naucalpan de Juárez, 53 950 Mexico

**Mongolia**
> Mongolian National Centre for Standardization and Metrology (MNCSM)
> P.O. Box 48, Ulaanbaatar 211051

**Morocco**
> Service de normalisation industrielle marocaine (SNIMA)
> Ministère do commerce, de l'industrie et l'artisanat
> Angle Avenue Al Filao, et Rue Dadi Secteur 21 Hay Riad, 10100 Rabat

**Netherlands**
> Nederlands Normalisatie-instituut (NNI)
> Kalfjeslaan 2, P.O. box 5059, NL-2600 GB Delft

**New Zealand**
> Standards New Zealand (SNZ)
> Radio New Zealand House, 155 the Terrace, Wellington 6001

**Nigeria**
> Standards Organisation of Nigeria (SON)
> Federal Secretariat, Phase 1, 9th Floor, Ikoyi, Lagos

**Norway**
> Norges Standardiseringsforbund (NSF)
> Drammensveien 145 A, Postboks 353 Skoyen, N-0212 Oslo

**Pakistan**
> Pakistan Standards Institution (PSI)
> 39 Garden Road, Saddar, Karachi-74400

**Panama**
> Comisión Panameña de Normas Industriales y Técnicas (COPANIT)
> Ministerio de Comercio e Industrias, Apartado Postal 9658, Panama, Zona 4

**Philippines**
> Bureau of Product Standards (BPS)
> Dept. of Trade and Industry, 361 Sen. Gil J. Puyat Avenue, Makati
> Metro Manila 1200

## Poland
Polish Committee for Standardization (PKN)
ul. Elektoralna 2, P.O. Box 411, PL-00-950 Warszawa

## Portugal
Insituto Português da Qualidade (IPQ)
Rua C à Avenida dos Três Vales, P-2825 Monte de Caparica

## Romania
Institutul Român de Standardizare (IRS)
Str. Jean-Louis Calderon Nr. 13, Cod 70201, R-Bucuresti 2

## Russian Federation
State Committee of the Russian Federation for Standardization, Metrology and
Certification (GOSTR)
Leninsky Prospekt 9, Moskva 117049

## Saudi Arabia
Saudi Arabian Standards Organization (SASO)
Imam Saud Bin Abdul Aziz Bin Mohammed Road (West End)
P.O. Box 3437, Riyadh 11471

## Singapore
Singapore Productivity and Standards Board (PSB)
1 Science Park Drive, Singapore 118221

## Slovakia
Slovak Office of Standards, Metrology and Testing (UNMS)
P.O. Box 76, Stefanovicova 3, 810 05 Bratislava

## Slovenia
Standards and Metrology Institute of the Republic of Slovenia (SMIS)
Kotnikova 6, SI-1000 Ljubljana

## South Africa
South African Bureau of Standards (SABS)
1 Dr Lategan Rd, Groenkloof, Private Bag X191, Pretoria 0001

## Spain
Asociación Española de Normalización y Certificación (AENOR)
Génova, 6, E-28004 Madrid

## Sri Lanka
Sri Lanka Standards Institution (SLSI)
53 Dharmapala Mawatha, P.O. Box 17, Colombo 3

## Sweden
Standardiseringen I Sverige (SIS)
St Eriksgatan 115, Box 6455, S-113 82 Stockholm

**Switzerland**
Swiss Association for Standardization (SNV)
Mûhlebachstrasse 54, CH-8008 Zurich

**Syrian Arab Republic**
Syrian Arab Organization for Standardization and Metrology (SASMO)
P.O. Box 11836, Damascus

**Tanzania**
Tanzania Bureau of Standards (TBS)
Ubungo Area, Morogoro Road/Sam Nujoma Road, Dar es Salaam

**Thailand**
Thai Industrial Standards Institute (TISI)
Ministry of Industry, Rama VI Street, Bangkok 10400

**Trinidad and Tobago**
Trinidad and Tobago Bureau of Standards (TTBS)
P.O. Box 467, Port of Spain

**Tunisia**
Institut national de la normalisation et de le propriété industrielle (INNORPI)
B.P. 23, 1012 Tunis-Belvédère

**Turkey**
Tûrk Standardlari Enstitûsû (TSE)
Necatibey Cad. 112, Bakanliklar, TR-06100 Ankara

**Ukraine**
State Committee of Ukraine for Standardization, Metrology and Certification (DSTU)
174 Gorky Street, GSP, Kyiv-6, 252650

**United Kingdom**
British Standards Institution (BSI)
389 Chiswick High Road, GB-London W4 4AL

**United States of America**
American National Standards Institute (ANSI)
11 West 42nd Street, 13th Floor, New York, NY 10036

**Uruguay**
Instituto Uruguayo do Normas Técnicas (UNIT)
Pza. Independencia 812, Piso 2, Montevideo

**Uzbekistan**
Uzbek State Centre for Standardization, Metrology and Certification (UZGOST)
Ulitsa Farobi, 333-A, 700049 Tachkent

**Venezuela**
Fondo para la Normalización y Certificación de la Calidad (FONDONORMA)
Av. Andrés Bello-Edf Torre Fondo Común, Pisos 11 y 12 - Apartado Postal
51116, Caracas 1050-A

Vietnam
>    Directorate for Standards and Quality (TCVN)
>    70, Tran Hung Dao Street, Hanoi

Yugoslavia
>    Savezni zavod za standardizaciju (SZS)
>    Kneza Milosa 20, Post Pregr. 933, YU-11000 Beograd

Yugoslav Republic of Macedonia, Former
>    Zavod za standardizacija I metrologija (ZSM)
>    Ministry of Economy, Samoilova 10, 91000 Skopje

Zimbabwe
>    Standards Association of Zimbabwe (SAZ)
>    P.O. Box 2259, Harare

# Index

# IMMIGRATION AT THE
# GOLDEN GATE